Mathematics in Early Childhood Education

Second edition

Mathematics in Early Childhood Education

Second edition

Amy MacDonald

OXFORD
UNIVERSITY PRESS

Oxford University Press is a department of the University of Oxford. It furthers the University's objective of excellence in research, scholarship and education by publishing worldwide. Oxford is a registered trademark of Oxford University Press in the UK and in certain other countries.

Published in Australia by
Oxford University Press
Level 8, 737 Bourke Street, Docklands, Victoria 3008, Australia.

First published 2018
2nd Edition 2024

ISBN 9780190338756

Edited by Bluepen Editorial Services
Typeset by Newgen KnowledgeWorks Pvt. Ltd., Chennai, India
Proofread by Amy Moore
Indexed by Puddingburn Publishing Services Pty. Ltd.
Printed in China by Leo Paper Products Ltd

Oxford University Press Australia & New Zealand is committed to sourcing paper responsibly.

Acknowledgement of Country
Oxford University Press acknowledges the Traditional Owners of the many lands on which we create and share our learning resources. We acknowledge the Traditional Owners as the original storytellers, teachers and students of this land we call Australia. We pay our respects to Elders, past and present, for the ways in which they have enabled the teachings of their rich cultures and knowledge systems to be shared for millennia.

Warning to First Nations Australians
Aboriginal and Torres Strait Islander peoples are advised that this publication may include images or names of people now deceased.

Contents

Extended contents

FIGURES

TABLES

OPINION PIECES

LEARNING EXPERIENCE PLANS

PREFACE

Allow me to introduce you to *Mathematics in Early Childhood Education*. I hope this book will become a trusted friend to you throughout your teacher education studies—and beyond. This book has been designed to be a comprehensive guide that encompasses a framework for early childhood mathematics education, approaches to mathematics teaching with young children, and content for mathematics teaching. This book is informed by, and presents, current Australasian and international research relevant to early childhood mathematics education. There are also many examples from the field—that is, real artefacts and anecdotes from children, families, pre-service teachers, and current educators.

The coverage of this text is inclusive of the full early childhood age range of birth to eight years, and is also inclusive of the variety of educators who work with children from this age range. The text explores mathematics in way that is relevant to a range of early childhood contexts, including home, playgroup, childcare, preschool, and school; and mathematics at the time of transition to school receives explicit treatment.

This text provides clear links to the two national curricula that inform early mathematics education in Australia: the *Early Years Learning Framework for Australia* (EYLF) and the *Australian Curriculum: Mathematics*. Links to curriculum outcomes are included as chapter features, and examples of how mathematical content can be mapped to both curriculum documents are presented. These examples show how mathematics learning can be 'bridged' across prior-to-school and school settings, and how the use of a common language can allow the seemingly disparate curriculum documents to communicate with one another.

This book encourages you to become a reflective mathematics educator. Throughout the text you will find reflections from current educators about their practice, and you are also invited to 'pause and reflect' on your own practice. Hopefully these provocations will help you to reflect upon, and better understand, yourself as a mathematics educator and how your mathematics education practices are shaped. You may also find that these invitations to reflect form the basis for interesting conversations with those around you.

There are a number of voices in this book, in addition to my own—voices of researchers, educators, teacher education students, and children. They are the voices of the people to whom I listen—the people who influence the ways I think about early childhood mathematics education. Some of these people research and teach in mathematics education specifically, but many don't. Indeed, many of the people you will meet in this book have expertise in areas other than mathematics education, but we can learn a great deal about how to be excellent early childhood mathematics educators by listening to them. I hope that by sharing their voices in this way, their opinions, experiences, and reflections may influence you and the ways in which you see yourself and your practice—just as they have influenced me.

LEARNING LINK

YOUR DIGITAL STUDENT RESOURCES

Keep an eye out for this icon throughout the text: . This indicates where additional digital resources, relevant to the section of text you are reading, are available.

Go to www.oup.com/he/macdonald2e.

Go to the inside cover of this book to find your code.

If you don't already have an Oxford Learning Link account

Go to https://learninglink.oup.com/register

Enter your information into the fields to create a student account, inputting your Access Code into the last field.

Access to the resources will be granted once you log in and will appear on your 'My Account' page.

If you already have an Oxford Learning Link account

Log in at: https://learninglink.oup.com/signin

Type the name of the ebook into a search box: *Mathematics in Early Childhood Education.*

On the results screen, click on the title of your ebook, then click on the 'Student Resources'.

In the 'Need access to "locked" resources' box, click on 'Click here'.

In the overlay window, follow the instructions for the centre option 'Redeem an access code'. On redeeming your Access Code, you'll be granted access to the resource for the duration of time denoted with the code.

If you have any issues

Please email cs.au@oup.com.

What are the students resources?

- Video summaries of the main point in each chapter
- Audio of the educator reflections and opinion pieces
- Flashcards of key terms with definitions
- End of chapter online quizzes
- Learning experience template.

GUIDED TOUR

Chapter overview: A brief outline of the focus and intended outcomes of each chapter.

Go online to watch a video of Paige Lee explain the key message in each chapter.

Learn more
Go online to see a video from Paige Lee explaining the key messages of this chapter.

CHAPTER OVERVIEW

This chapter will provide an overview of early childhood mathematics education and will justify the importance of exploring mathematics in the early childhood years. The chapter begins by outlining the body of research that provides evidence for young children's mathematical competence. It then considers what might be meant by 'early childhood mathematics education' and concludes with a useful framework for exploring mathematics in the everyday lives of young children.

In this chapter, you will learn about the following topics:

» views of mathematics
» mathematics and numeracy
» young children's mathematical competence

KEY TERMS

» Early childhood
» Early childhood mathematics education
» Mathematics
» Noticing

Key terms: Succinct explanations of mathematics, science and technology concepts and other key terms. They are listed at the beginning of the chapter, defined in the margins where they first appear and collated in the Glossary at the end of the book.

Pause and reflect: Critical reflection questions that encourage you to reflect upon your own experiences, consider issues and interpret examples.

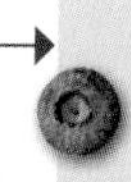

Pause and reflect

What is *your* view of mathematics? Do you see it as 'universal truth', 'socio-cultural practice', or a combination of the two? How does your view influence your ideas about mathematics teaching and learning?

Educator reflection: Narratives from early childhood educators that illustrate their work with young children engaging with mathematics.

Go online to listen to the Educator Reflections. Some have been read by the authors.

Educator reflection

As a child I remember not liking maths at all! Funny, I remember my mum not liking maths also … Maths to me was dull; it always had to be written and was difficult. This was the only subject I ever really had trouble with, so obviously I wasn't confident at it. However, I also have a memory of when I was in Grade 2, so I would have been seven years old. My teacher during a whole group time had different-sized containers

Learn more
Go online to hear Keira Hood read this Educator reflection.

Opinion Piece 6.1

ASSOCIATE PROFESSOR WENDY GOFF

I believe that when the adults in the lives of young children come together to support mathematical learning, a variety of opportunities emerge. Coming together provides an avenue for questioning and learning about mathematics, and also an avenue for sharing and learning about the mathematical understandings that children are

Learn more
Go online to hear Associate Professor Wendy Goff read this Opinion Piece.

Opinion Piece: Comments from educators about why they enjoy teaching mathematics to young children.

Go online to listen to Opinion Pieces. Some have been read by the authors.

CURRICULUM CONNECTIONS

Early Years Learning Framework

Outcome 1: Children have a strong sense of identity

Children develop their emerging autonomy, inter-dependence, resilience and sense of agency.

This is evident when children:

Curriculum connections: Links that draw attention to the EYLF and how educators contribute to outcomes.

STRENGTHS	IDEAS FOR LEARNING	MATHEMATICAL CONCEPTS
Interested in dinosaurs	Setting up several of his favourite dinosaurs in height order. Drawing his favourite dinosaurs in height order or heaviest to lightest, etc.	Sorting by size Size, area
Playing with his sister and dressing up as a superhero	Superhero game: Dress up as superhero and his sister has to hide, before the superhero finds her—he has to count backwards from 10 to zero before blast off. Hide some of the child's toys around the house and the superhero needs to find them.	Number order Counting on—how many has he found, how many left? Subtraction—how many more does he need to find?

Learning Experience Plan: Samples of mathematical learning experiences collected from pre-service teachers and early childhood educators that you can use, adapt or revise to your expected teaching needs and lesson planning.

For further reflection: Questions to encourage you to reflect upon the concepts discussed throughout the chapter.

FOR FURTHER REFLECTION

1 Can you recall a time when you felt excited by, and confident with, mathematics?
2 What were the factors that enabled you to feel confident as a mathematician?
3 How can you apply this to your own mathematics teaching practices to help the children with whom you work feel confident and excited?

Further reading: Recommendations that allow you to access additional information about the themes being explored in each chapter.

FURTHER READING

Australian Association of Mathematics Teachers, & Early Childhood Australia. (2006). *Position paper on early childhood mathematics*. Adelaide, SA & Watson, ACT: Authors.

Geist, E. (2009). Children and mathematics: A natural combination. In *Children are born mathematicians: Supporting mathematical development, birth to age 8* (pp. 1–34). Upper Saddle River, NJ: Pearson Education.

Jorgensen, R., & Dole, S. (2011). The changing face of school mathematics. In *Teaching mathematics in primary schools* (pp. 1–22). Crows Nest, NSW: Allen & Unwin.

ABOUT THE AUTHOR

Associate Professor Amy MacDonald is an Associate Professor of Early Childhood Mathematics Education and Associate Head of School in the School of Education at Charles Sturt University, Albury–Wodonga, Australia. Amy has a Bachelor of Education (Primary) (Honours Class 1) and a Doctor of Philosophy in mathematics education, both from Charles Sturt University. She has published a large number of books, book chapters, journal articles, and conference papers in the areas of mathematics education, early childhood education, and educational transitions. Amy has received several awards for her work, including the Mathematics Education Research Group of Australasia Early Career Award, the New South Wales Institute for Educational Research Beth Southwell Research Award, and the Charles Sturt University Outstanding Thesis Prize for her PhD. Amy taught mathematics, science and technology education in Charles Sturt University's Bachelor of Education (Birth to Five Years) program for more than a decade, and in 2014 she was awarded an Australian Government Office for Learning and Teaching 'Citation for Outstanding Contribution to Student Learning' for her approaches to mathematics education.

ACKNOWLEDGMENTS

I would like to thank all of the people who have inspired and influenced this book. First, I would like to thank my husband Cody and son Jarvis for their help, support, and patience during the writing of *Mathematics in Early Childhood Education*. As you read this book, you will come to know a little about my Jarvis—and hopefully you will see how much inspiration I have drawn from him.

I would like to offer sincere thanks to Geraldine Corridon from Oxford University Press for the opportunity to produce a second edition of this book. I deeply value my ongoing relationship with OUP and the terrific team of people who work hard to produce textbooks that are meaningful and relevant to their audiences.

My thanks also go to the wonderful Paige Lee, who provided extensive feedback on the first edition of this text and has been instrumental in the production of this second edition. Paige has generously contributed a number of images and examples, and has worked closely with me to update the reference literature. Thank you, Paige.

Extra special thanks are extended to everyone who contributed anecdotes, reflections, photographs, work samples, learning stories, learning experience plans, and opinion pieces to this book. These contributions are the backbone of the book, and I am grateful for the generosity of the following people:

Julia Alexander, Penny Baker, Sarah Barcala, Sussann Beer, Elizabeth Bowden, Michelle Call, Fiona Collins, Sharon Cope, Jessamy Davies, Sue Dockett, Sandra Dos Reis, Sheena Elwick, Angela Fenton, Aimie Gibson, Jason Goldsmith, Michelle Gunter, Lynette Hartley, Stephanie Hill, Keira Hood, Kathryn Hopps, Varinder Kaur, Paige Lee, Nikki Masters, Sarah Morrow, Michelle Muller, Rebecca O'Gorman, Vicki Olds, Kay Owens, Maree Parkes, Kate Pearce, Bob Perry, Gabrielle Pritchard, Alexandra Roth, Jody Rumble, Patsy Saul, Samira Smith, Valerie Tillett, Amy Urquhart, Cen (Audrey) Wang, Stephanie Pappas, Patricia Woodford.

Finally, thank you to every teacher education student with whom I have had the pleasure of working and to every student I am yet to meet—this book is for you.

Part 1

FRAMING THE TEACHING OF MATHEMATICS IN EARLY CHILDHOOD EDUCATION

OVERVIEW OF PART 1

Part 1 of this book consists of four chapters that, taken as a whole, provide an overarching framework for the teaching of mathematics in early childhood education. Individually, the chapters address the following topics:

» introduction to early childhood mathematics education

» becoming an early childhood mathematics educator

» using early childhood theory to inform mathematics education

» early childhood mathematics curricula.

Collectively, these chapters provide different lenses on early childhood mathematics education, including theoretical perspectives, the role of the numerate educator, and the place of mathematics in various early childhood curricula. It is anticipated that Part 1 will provide a useful framework for interpreting the subsequent parts and chapters of this book, and it is likely that you will revisit these opening chapters a number of times as you progress through the book. Indeed, this book purposefully unfolds layers of knowledge relevant to early childhood mathematics education—chapter by chapter, part by part—to help you in building up a comprehensive professional identity as a mathematics educator that is informed by a range of different but complementary perspectives.

01 Introduction

Learn more
Go online to see a video from Paige Lee explaining the key messages of this chapter.

CHAPTER OVERVIEW

This chapter will provide an overview of early childhood mathematics education and will justify the importance of exploring mathematics in the early childhood years. The chapter begins by outlining the body of research that provides evidence for young children's mathematical competence. It then considers what might be meant by 'early childhood mathematics education' and concludes with a useful framework for exploring mathematics in the everyday lives of young children.

In this chapter, you will learn about the following topics:

- views of mathematics
- mathematics and numeracy
- early childhood mathematics education
- young children's mathematical competence
- noticing, exploring, and talking about mathematics.

KEY TERMS

- Early childhood
- Early childhood mathematics education
- Mathematics
- Noticing
- Numeracy

Introduction

There is growing recognition that children explore a range of mathematical ideas from a very young age. Researchers used to think that very young children have very little knowledge of, or capacity to learn, mathematics; however, contemporary research argues that mathematical competencies are either innate or develop in the first years of life (Sarama & Clements, 2009). Furthermore, it has been suggested that the increasing numbers of children participating in early childhood programs, and the growing recognition of the importance of mathematics in general, provide compelling reasons for understanding children's mathematical development in the early childhood years (Doig, McRae, & Rowe, 2003). In this first chapter, I lay the foundations for the importance of mathematics education in the early childhood years. This book adopts the definition of **early childhood** as being the period from birth to eight years of age and considers children's learning within and across a range of settings, including early childhood education and care, early primary school, and home and community contexts. With this in mind, in this book **early childhood mathematics education** can be taken to mean the opportunities for learning about **mathematics**, both formally and informally, across the range of contexts in which young children participate. Both early childhood and school settings are considered, as are the specific curricula that relate to these settings. However, the book is written with the intention of being relevant to educators across both of these settings without being specific to one or the other. Indeed, it is important to consider mathematics education across the transition to school in terms of both continuity *and* change; hence, the content of this book might be relevant to prior-to-school settings *or* school settings *or* both equally and at the same time.

Early childhood is defined as the period from birth to eight years of age.

Early childhood mathematics education refers to the opportunities for learning about mathematics across the range of contexts in which young children participate.

Mathematics is a knowledge domain that is about patterns, relationships, representations, symbols, abstraction, and generalisations.

Throughout this book, 'Curriculum Connections' features provide links between chapter content and the two curricula guiding early children's mathematics education in Australia: *Belonging, Being and Becoming: The Early Years Learning Framework for Australia* (EYLF; Australian Government Department of Education and Training [DET], 2019); and the *Australian Curriculum: Mathematics* (Australian Curriculum, Assessment and Reporting Authority [ACARA], 2022). These features are intended as a guide for how the mathematical content and pedagogical approaches explored in this text align to the curricula with which early years educators work. It is anticipated that these links will assist you to engage with curricula in meaningful ways and will serve as a model for linking your own pedagogical practices to appropriate curricula and other frameworks that guide your teaching practice.

Views of mathematics

'Mathematics' is a term that may be taken for granted; however, it is usually the case that people have differing definitions of—and differing views of what constitutes—mathematics. It is likely that people's individual experiences of mathematics will in large measure account for these differences. Some people view mathematics as a body of 'universal truths' that teachers transfer or impart to their students as a set of facts and skills; while others view mathematics as a socio-cultural practice that is a product of reflective human activity (Siemon et al., 2011).

Our views of what constitutes mathematics shape our decisions about how we teach and learn mathematics (Siemon et al., 2011):

> If mathematics is viewed as a set of universal truths, teachers are more likely to see their task as transferring a given set of facts and skills to students and to view

> student learning as the capacity to reproduce these facts and skills as instructed. If, on the other hand, mathematics is viewed as a socio-cultural practice, a product of reflective human activity, then it is more likely that teachers will see their task as engaging students in meaningful mathematical practices and view student learning in terms of conceptual change (p. 6).

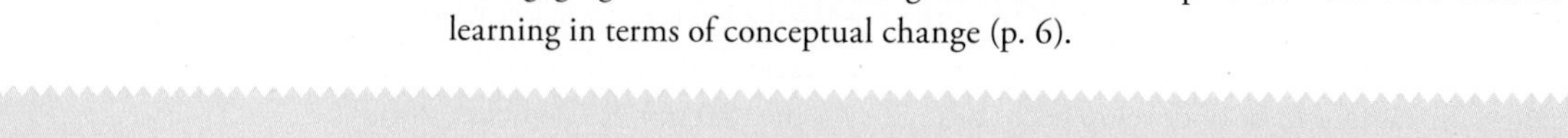

Pause and reflect

What is *your* view of mathematics? Do you see it as 'universal truth', 'socio-cultural practice', or a combination of the two? How does your view influence your ideas about mathematics teaching and learning?

Across the contemporary mathematics education literature and curriculum documents, a typical definition is that mathematics is a discipline or knowledge domain that 'is about seeking patterns and relationships, representing these, symbolising these ideas, and eventually learning to abstract and generalise' (Bobis, Mulligan, & Lowrie, 2013, p. 6). In a similar fashion, Jorgensen and Dole (2011) describe mathematics as the study of patterns and relationships; a way of thinking, seeing and organising the world; a language; a tool; a form of art; and power. It is important to recognise that mathematical ideas have evolved across all cultures over thousands of years, and are constantly developing (ACARA, 2022).

Mathematics and numeracy

Reflecting the influence of socio-cultural perspectives of mathematics, a common discourse in modern mathematics education contexts is articulated around the notion of 'numeracy'. It is sometimes the case that the terms 'mathematics' and 'numeracy' are presented as an either/or proposition—positioned as separate from one another. However, it is perhaps more useful to think of the two as being inextricably interconnected, with each building upon and informing the other.

Numeracy is a social and cultural perspective for discovering, thinking about, and applying mathematical knowledge.

Starting from a view of mathematics as a discipline or knowledge domain, the concept of **numeracy** provides 'a social and cultural perspective for discovering and thinking about mathematical knowledge and applying it to fulfil the purposes of our everyday lives' (Macmillan, 2009, p. 1). Numeracy emphasises:

- the context, purpose, and usefulness of a particular approach in solving problems in everyday life
- flexible, negotiable, and meaningful applications of mathematical concepts
- processes of applying mathematical concepts or operations
- appreciation of the mathematical dimensions of everyday experiences
- the use of available knowledge, skills, intuition, creativity, experiences, resources, and tools
- the development of confidence alongside competence.

In short, the concept of numeracy provides a useful means for thinking about the *how* and *why* of mathematics education.

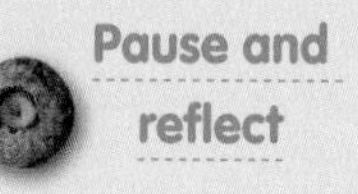

Pause and reflect

Have a go at writing your own definitions of *mathematics* and *numeracy*. How does this help you to understand the relationship, and distinction, between the two?

Early childhood mathematics education

Over the years, there have been competing views as to whether mathematics can, or *should*, be explored with young children. Influential pedagogues in early childhood education—including Friedrich Froebel and Maria Montessori—argued that young children are capable of complex mathematical thought and enjoy using mathematics to explore and understand their world (Balfanz, 1999). In contrast, a number of educators, psychologists, and social theorists have argued that it is inappropriate and unnecessary to introduce young children to mathematics education in an organised fashion (Balfanz, 1999). Indeed, these tensions are still evident—we are exposed to varying interpretations of discourses such as 'let the children play' and 'play as powerful learning'.

However, a significant body of research provides compelling evidence that young children *can* and *do* explore complex mathematical ideas as part of their everyday lives. Graue et al. (2015) suggest it is critical for teachers to see themselves as more than a Piagetian facilitator who steps back after setting up the children's learning environments. Rather, they should view mathematics teaching from a Vygotskian perspective, as a scaffolding role, building on the children's funds of knowledge to maximise the mathematics *within* children's play without *disrupting* their play (MacDonald & McGrath, 2022). As Geist (2009, p. 12) says, we should 'think of children as competent mathematicians'. Contemporary early childhood practice advocates for the provision of early childhood mathematics education, facilitated by educators who have deep conceptual knowledge and skills in identifying and building upon the mathematical possibilities in children's everyday activities and play. De Vries, Thomas, and Warren (2010) argue for early childhood educators to 'focus on how [they] construct themselves as teachers engaged in both play-based pedagogy and mathematics as a curriculum discipline' (p. 722). Educators working with very young children need to be willing to include mathematics education in their programming and to look for opportunities to emphasise or include mathematics in an engaging manner wherever possible with children (MacDonald & McGrath, 2022).

In 2006, the Australian Association of Mathematics Teachers (AAMT) (the Australian peak body for mathematics education) and Early Childhood Australia (ECA) (the Australian peak body for early childhood education) published a joint *Position Paper on Early Childhood Mathematics*. This seminal document provided a compelling rationale for the provision of high-quality early childhood mathematics education and recommended appropriate actions to ensure that all young children have access to powerful mathematical learning that nurtures success and positive dispositions (AAMT & ECA, 2006). The following is the full list of recommendations for early childhood educators.

POSITION PAPER ON EARLY CHILDHOOD MATHEMATICS: RECOMMENDATIONS FOR EARLY CHILDHOOD EDUCATORS

Early childhood educators should adopt pedagogical practices that:

- engage the natural curiosity of young children to assist in the development of the children's mathematical ideas and understandings

- use accepted approaches to early childhood education such as play, emergent, child-centred and child-initiated curriculum to assist young children's development of mathematical ideas
- ensure that the mathematical ideas with which young children interact are relevant to their present lives as well as forming the foundation for future mathematical learning
- recognise, celebrate and build upon the mathematical learning that young children have developed and use the children's methods for solving mathematical problems as the basis for future development
- encourage young children to see themselves as mathematicians by stimulating their interest and ability in problem solving and investigation through relevant, challenging, sustained and supported activities
- recognise that mathematical learning is a social activity supported and extended through interaction with both other children and adults
- provide appropriate materials, space, time and other resources to encourage children to engage in their mathematical learning
- focus on the use of language to describe and explain mathematical ideas, recognising the important role language plays in the development of all learning
- address the learning needs of children with intellectual disabilities through explicit teaching of applicable vocabulary and other strategies that are appropriate for each child
- attend to the language learning needs in mathematics of children for whom English is a second or subsequent language
- respond to the diverse cultural backgrounds of young children in this country and ensure that all children, particularly those from more traditional Indigenous communities, have access to cultural and language learning that underpins learning of western mathematics
- encourage young children to justify their mathematical ideas through the communication of these ideas in ways devised by the children that display appropriate levels of mathematical rigour
- acknowledge that while materials may be important in young children's development of mathematical ideas, these ideas are actually developed through thinking about action—children need to be encouraged to engage in mental manipulation of mathematical ideas
- recognise that children's mathematical development occurs within, is affected by, and needs to be relevant to a number of different contexts including family, cultural groups, community, prior-to-school setting and school
- assess young children's mathematical development through means such as observations, learning stories, discussions etc. that are sensitive to the general development of the children, their mathematical development, their cultural and linguistic backgrounds, and the nature of mathematics as an investigative, problem solving and sustained endeavour

- recognise that the primary use for gathering information about children's mathematical development through assessment is to track that development and to help plan further interactions, tasks, activities and interventions.

(Source: Australian Association of Mathematics Teachers & Early Childhood Australia, 2006)

These views are reflected in the two national curricula of relevance to Australian early childhood education settings: the EYLF and the *Australian Curriculum: Mathematics*. The EYLF (DET, 2019) states:

> The Framework forms the foundation for ensuring that children in all early childhood education and care settings experience *quality teaching and learning*. It has a specific emphasis on *play-based learning* and recognises the importance of communication and language (including early literacy and *numeracy*) and social and emotional development (p. 5).

*Emphasis added

This is built upon by the *Australian Curriculum: Mathematics*:

> The early years (5–8 years of age) lay the foundation for learning mathematics. Students at this level *can access powerful mathematical ideas that are relevant to their current lives*, and that it is the relevance to them of this learning that prepares them for the following years. Learning the language of mathematics is vital in these early years (National Curriculum Board, 2009, p. 7).

*Emphasis added

The specific mathematics content and pedagogies associated with the EYLF and the *Australian Curriculum: Mathematics* will be discussed further in Chapter 4, and explicit links to these curricula will be made throughout the chapters of this book.

Young children's mathematical competence

Children begin developing mathematical skills from a very young age. International research has shown that babies and toddlers demonstrate competence in regard to a range of mathematical concepts and processes, including number and counting, space and geometry, measurement, dimensions and proportions, location, pattern, classification, and problem solving (Björklund, 2008; Lee, 2012; Reikerås, Løge, & Knivsberg, 2012). Both Australian and international research has established that young children engage with a range of mathematical concepts and processes prior to starting school (for example, Gervasoni & Perry, 2015; Sarama & Clements, 2015). The seminal Australian study, the *Early Numeracy Research Project* (see, for example, Clarke, Clarke, & Cheeseman, 2006), investigated the mathematical knowledge of over 1400 children in their first year of primary school. An important finding from the study was that much of the content that formed the mathematics curriculum for the first year of school was already understood clearly by many children on arrival at primary school (Clarke et al., 2006), a finding echoed in several other studies, both in Australia (for example, Gervasoni & Perry, 2015; MacDonald, 2010) and internationally (for example, Aubrey, 1993; Wright, 1994). Research has emphasised the importance of this early mathematical learning, with links being drawn between early mathematics and later achievement (MacDonald & Carmichael, 2016; Watts, Duncan, Siegler, & Davis-Kean, 2014). This research notes, in particular, the predictive power of mathematical knowledge at school entry for later mathematical achievement (Duncan et al., 2007). It has been found that children who enter primary school with high levels of mathematical knowledge maintain these high levels of mathematical skill throughout, at least, their primary school education (Baroody, 2000; Klibanoff, 2006).

Furthermore, De Lange (2008) has suggested that in the years prior to commencing formal education, young children have a curiosity about scientific phenomena, including mathematics. In an Australian study of teacher-reported data for 6500 children, MacDonald and Carmichael (2015) found that 98% of the children showed interest in numbers at 4–5 years. If children engage in meaningful and enjoyable mathematics education in the early childhood years, they are much more likely to appreciate and continue to engage in later mathematics education (Linder, Powers-Costello, & Stegelin, 2011).

Of course, there will be substantial variance in the mathematical competencies children develop prior to school (Peter-Koop & Kollhoff, 2015), and both standardised tests and experimental tasks reveal marked individual differences in children's mathematical knowledge by the time children enter preschool (Levine et al., 2010). Given the compelling research pertaining to the relationship between mathematics at the time of school entry and later school achievement, it is important to consider the mathematical competencies of children in the early years in order to understand the foundation on which subsequent mathematics education should build.

Educator reflection

Learn more
Go online to hear this Educator reflection being read.

I currently work in the toddler room but that doesn't stop me from using critical and important words when working alongside children. I believe it's important to expose young children—even those at age two—to the rich describing words. For example, when climbing on the large equipment outside I will always talk as the children are making their way 'along' the plank and 'over' the bar and then 'down' the slide. Simple words like these encourage children to think deeper about what it is that they are doing to further enhance spatial awareness.

(Source: Amy Urquhart)

Noticing, exploring, and talking about mathematics

Children's everyday lives provide many opportunities to engage with mathematical concepts and processes through play, exploration, routines, and activities. Part of an educator's role in an early childhood setting is to notice what children know and are learning about mathematics, how they come to understand this mathematics, and how they then put this knowledge into practice (Marcus, Perry, Dockett, & MacDonald, 2016). Mason (2002, p. 29) describes **noticing** as 'a collection of practices both for living in, and hence learning from, experience, and for informing future practice'. Moreover (Mason, 2002):

Noticing
is a collection of practices for living in, and learning from, experience.

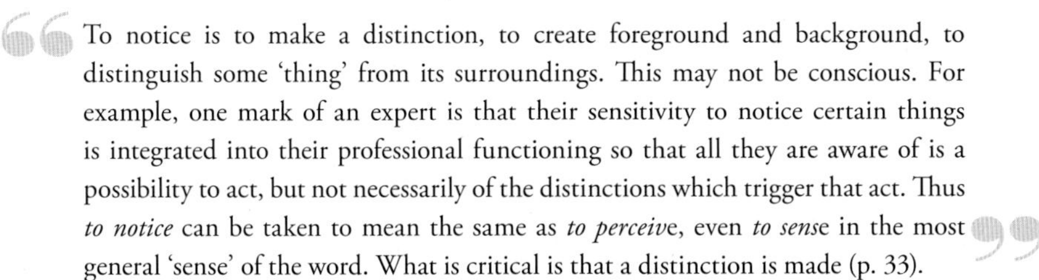
To notice is to make a distinction, to create foreground and background, to distinguish some 'thing' from its surroundings. This may not be conscious. For example, one mark of an expert is that their sensitivity to notice certain things is integrated into their professional functioning so that all they are aware of is a possibility to act, but not necessarily of the distinctions which trigger that act. Thus *to notice* can be taken to mean the same as *to perceive*, even *to sense* in the most general 'sense' of the word. What is critical is that a distinction is made (p. 33).

Building on the work of Mason, the early mathematics program *Let's Count* (Gervasoni & Perry, 2015; Perry & Gervasoni, 2012) advocates a framework of *noticing*, *exploring*, and *talking about* mathematics as a way of recognising, and building upon, the mathematics with which children engage in everyday contexts.

Opinion Piece 1.1

PROFESSOR BOB PERRY

Learn more
Go online to listen to this Opinion Piece.

I recently had the privilege to visit some Aboriginal rock paintings in Far North Queensland. One cannot fail to be impressed by the age, the extent, and the artistry of these paintings. Of course, the paintings are an important part of both past and current Aboriginal culture and tell stories that will help ensure that the cultures continue. While I appreciate and understand all of this, what did I notice about the paintings? I noticed how many there were in the particular gallery—I didn't count them but noticed that there were more than twenty. I noticed size—some of the paintings were 'life-size', some were smaller, and one was 'enormous'. I noticed symmetry, I noticed position, I noticed symbolism. In short, I noticed mathematics. Perhaps this is inevitable after forty-five years' thinking about mathematics and mathematics education. Perhaps it is because there is 'mathematics in everything'.

It is important that teachers and families 'notice' their children's mathematics as this is how they can build knowledge about what the child can do, how the child does whatever it is, and what the teacher or family member might introduce to the child next. As well, it makes children feel pretty important to have adults noticing what they are doing.

However, 'noticing' is only the start. Children need to be given opportunities to explore and enhance the mathematics in which they are involved and they will do that if the adults provide opportunities for exploration and then encourage them to talk about the mathematics they find. So, notice, explore, and talk about children's mathematics and you will make a difference to lots of children.

Bob Perry is recently retired from forty-five years of university work, is Professor Emeritus at Charles Sturt University, Albury–Wodonga, Australia, and Director, Peridot Education Pty Ltd. Bob's current research interests include powerful mathematics ideas in preschool and the first years of school; noticing mathematics; transition to school, with particular emphasis on starting school within families with complex support needs; preschool education in remote and Indigenous communities, and evaluation of educational programs. Bob and partner, Sue Dockett, have researched and published in the area of transitions to school. This work is internationally and nationally renowned.

NOTICING MATHEMATICS

Young children explore mathematical ideas all the time—but it may be the case that the *mathematics* in these explorations goes unnoticed. For example, have you noticed that many children seem to enjoy 'covering' things? (Figure 1.1). Children might spread their food out so as to cover the entire plate, arrange books or toys to cover the top of the table, or smear paint all over the paper so that there is no white space left. When we look through a mathematical lens, this sort of investigatory play might be seen as children beginning to explore mathematical concepts such as area, position, and direction. When we start to notice these things, we notice other opportunities that children take to explore mathematical ideas—for example, burying items in the sandpit, covering their hands with paint, spreading jam on a piece of bread.

FIGURE 1.1 Noticing mathematics

(Source: Amy MacDonald)

Pause and reflect

What do *you* notice as children do these sorts of things? What do the *children* notice? What do you think they are discovering about shapes, spaces, and surfaces as they do these things?

EXPLORING MATHEMATICS

Once we *notice* the mathematics with which children engage, it is possible for us to *explore* it further. Look for the simple, everyday opportunities to explore mathematics with young children—they are often the most powerful learning opportunities. For instance, when working with infants, watch for opportunities that they take to navigate space, such as crawling under a table or through a tunnel (Figure 1.2), playing with a climbing frame, crawling over pillows, pulling themselves up to stand at tables or chairs. What do you notice as they perform these sorts of activity? What do you think they are discovering about the shapes and spaces around them? Activities such as these give infants the opportunity to explore shapes and spaces in very concrete ways, such as recognising openings, experiencing the length of objects as they crawl through or under them, and exploring the characteristics of the shapes and objects.

FIGURE 1.2 Exploring mathematics

(Source: Paige Lee)

Think about other play opportunities that might build on these types of exploration—for example, if you have noticed an infant who likes to climb *over* things, what other obstacles of different shapes and sizes can you provide for them to explore?

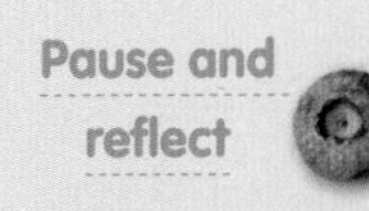

TALKING ABOUT MATHEMATICS

As you pursue mathematical ideas together, think about how a few careful questions or comments might help to extend children's understanding. There are lots of everyday opportunities for *talking about* mathematics. For example, eating fruit together is a great stimulus for mathematical conversation (Figure 1.3). Here are some suggestions for talking about mathematics as you and the children enjoy the fruit:

- What shapes can you find?
- What is similar/different about the pieces of fruit?
- Which is the largest/smallest piece? How do you know?
- How many pieces of fruit do we have? How many children would like to share this fruit? How many pieces will we have each?
- Let's eat the fruit in a pattern!

FIGURE 1.3 Talking about mathematics

(Source: Michelle Muller)

It is important to keep these sorts of conversations as natural and playful as possible. It is also important to look for other ways to consolidate the ideas you talk about—other opportunities to count, share, compare, sort, group, and create patterns.

Pause and reflect

Think about how you might use similar talking points during other everyday activities—for example, when packing away toys or the children's belongings. Also think about the mathematics that might be noticed, explored, and talked about during meal times or other routines, too.

In short, the *notice, explore, talk about* framework provides a useful means of conceptualising early childhood mathematics education. This framework will be revisited in later chapters of this book.

Chapter summary

In this chapter I have outlined the foundations for early childhood mathematics education, drawing upon the research evidence of young children's mathematical competence and linking this to a framework for exploring mathematics with young children. I extend to you an invitation to *notice*, *explore*, and *talk about* mathematics when you are spending time with children. Think about the simple, everyday activities children enjoy and how mathematical ideas might be explored. By doing so, you have the opportunity to make mathematics *meaningful* for children—and this is a very powerful thing indeed.

The specific topics we covered in this chapter were:

- views of mathematics
- mathematics and numeracy
- early childhood mathematics education
- young children's mathematical competence
- noticing, exploring, and talking about mathematics.

FOR FURTHER REFLECTION

Consider the image in Figure 1.4.

FIGURE 1.4 Shopping

(Source: Amy MacDonald)

1 What opportunities for *noticing*, *exploring*, and *talking about* mathematics might an everyday experience like shopping present for a young child?
2 How do these everyday experiences contribute to children's development of mathematical competencies?
3 How might numeracy be developed in this sort of experience?
4 How can these sorts of everyday experience be recognised and built upon in formal early childhood mathematics education contexts, such as in the early childhood centre or at school?

FURTHER READING

Australian Association of Mathematics Teachers, & Early Childhood Australia. (2006). *Position paper on early childhood mathematics*. Adelaide, SA & Watson, ACT: Authors.

Geist, E. (2009). Children and mathematics: A natural combination. In *Children are born mathematicians: Supporting mathematical development, birth to age 8* (pp. 1–34). Upper Saddle River, NJ: Pearson Education.

Jorgensen, R., & Dole, S. (2011). The changing face of school mathematics. In *Teaching mathematics in primary schools* (pp. 1–22). Crows Nest, NSW: Allen & Unwin.

MacDonald, A., & McGrath, S. (2022). Early childhood educators' beliefs about mathematics education for children under three years of age. *International Journal of Early Years Education*. https://doi.org/10.1080/09669760.2022.2107493

Macmillan, A. (2009). Introduction: Towards an inclusive philosophy and practice for early childhood numeracy education. In *Numeracy in early childhood: Shared contexts for teaching and learning* (pp. 1–19). South Melbourne, VIC: Oxford University Press.

Mason, J. (2002). Forms of noticing. In *Researching your own practice: The discipline of noticing* (pp. 29–38). London: RoutledgeFalmer.

Sarama, J., & Clements, D.H. (2009). Early childhood mathematics learning. In *Early childhood mathematics education research: Learning trajectories for young children* (pp. 3–25). London: Routledge.

02 Becoming an Early Childhood Mathematics Educator

Learn more
Go online to see a video from Paige Lee explaining the key messages of this chapter.

CHAPTER OVERVIEW

This chapter will explore the ways in which early years educators may come to see themselves as *mathematics* educators. The chapter focuses on the *affective* domain of mathematics, including how educators' own experiences with, and beliefs about, mathematics impact their approaches with children. The aim of this chapter is to help you understand your own mathematical identity and to see yourself as a teacher of mathematics. By building your own confidence and content knowledge in mathematics, you will be well positioned to sustain the mathematical confidence and curiosity of children.

In this chapter, you will learn about the following topics:

» the affective domain of mathematics education
» beliefs
» attitudes
» emotions
» mathematical identity
» memories of learning mathematics
» mathematical mindsets.

KEY TERMS

» Affective domain
» Attitudes
» Beliefs
» Dispositions
» Emotions
» Growth mindset
» Mathematical disposition
» Mathematical identity
» Mathematical mindset
» Mathematics anxiety

AMY MACDONALD

Introduction

In order to facilitate young children's mathematical development, it is important for educators to have thorough mathematical content knowledge and have *confidence* in using that knowledge. Moreover, early childhood mathematics education requires educators who understand the *affective* domain of mathematics, including how their own experiences with, and beliefs about, mathematics impact their approaches with children.

Mathematical dispositions and self-concepts influence educators' practices as much as they do children's learning of mathematics. In this chapter, a number of current early childhood educators share their memories of learning mathematics and reflect on how this has influenced their approaches to mathematics education. These stories demonstrate the power of our experiences; but also demonstrate how reflective practice enables us to understand *why* we might respond to things in particular ways, and what we might do to alter our responses for the benefit of children's mathematical learning.

The aim of this chapter is to help you understand your own mathematical identity and to see yourself as a teacher of mathematics. By building your own confidence and content knowledge in mathematics, you will be well positioned to sustain the mathematical confidence and curiosity of children. In short, the *more you know*, the *more you notice*—and as outlined in the opening chapter of this book, 'noticing' is crucial if we are to provide meaningful, interesting, challenging, and purposeful mathematical learning experiences for children (MacDonald, 2015a).

Opinion Piece 2.1

Learn more
Go online to listen to this Opinion Piece.

FIONA COLLINS

I have always loved mathematics, so becoming a mathematics educator was an easy and somewhat natural progression for me. However, that is not always the case for everyone.

In my profession, I am lucky enough to work with pre-service teachers while they are undergoing their training. Some come in and have a real passion for mathematics from the start, while others are non-committal or even apprehensive about becoming mathematics educators.

My greatest job satisfaction is realised when these pre-service teachers engage in mathematical learning and activities and I see a shift in their attitude and/or mannerisms. I see it on their faces, hear it in their voices and observe it in their teaching. They realise that maths can be fun and that they actually enjoy teaching it. They appreciate what it means to be a mathematics educator. It is this enthusiasm and joy that I wish I could bottle and give to every mathematics classroom around the globe.

Fiona Collins is a Lecturer in Primary Mathematics at Charles Sturt University, Wagga Wagga, Australia. She has extensive experience as a primary and secondary mathematics teacher in both metropolitan and regional schools in New South Wales. She is passionate about developing and nurturing a love of mathematics in people of all ages and is dedicated to sharing her knowledge, skills, and experience with others.

The affective domain of mathematics education

The **affective domain** of mathematics education refers to individuals' beliefs, attitudes, and emotions about mathematics (Grootenboer & Marshman, 2016). The affective domain plays a significant role in individuals' interest in, and response to, mathematics in general, and their use of mathematics in everyday life (Organisation for Economic Co-operation and Development, 2013). This response to mathematics is often described as our **mathematical disposition**—that is, our inclination or tendency to use mathematics.

Affective domain
refers to beliefs, attitudes, and emotions about mathematics.

Mathematical disposition
is the inclination or tendency to use mathematics.

BELIEFS

Philipp (2007, p. 259) defines **beliefs** as 'psychologically held understandings, premises, or propositions about the world that are thought to be true'. All educators hold beliefs: about their work, their children, their roles and responsibilities, and the curriculum areas they teach (Pajares, 1992; Howard, 2001). Mathematics is a curriculum area in which educators often hold strong beliefs about mathematics itself, and about themselves as learners and teachers of mathematics (Howard, 2004; Southwell & Khamis, 1994). Ernest (1989) outlined three views of mathematics that have implications for individuals' beliefs about mathematics. Table 2.1 presents Ernest's categories as summarised in Siemon et al. (2011, p. 26).

Beliefs
are psychologically held understandings about mathematics.

TABLE 2.1 Ernest's (1989) views of mathematics

INSTRUMENTALIST	Mathematics is seen as a useful kit of tools or techniques for use in everyday life. Learning mathematics is about acquiring practical facts and skills.
PLATONIST	Less interested in the uses of mathematics; rather, emphasis is on the relationships between mathematical ideas and the hierarchical structure of the discipline. Mathematics is regarded as existing independently of humans and awaiting discovery. Learning mathematics is about coming to understand the interconnectedness of mathematics and developing an appreciation of the discipline for its own sake.
PROBLEM-SOLVING	Mathematics is regarded as a dynamic and creative human invention, a process rather than a product. To learn mathematics is to engage in a creative process.

(Source: Siemon et al., 2011, p. 26)

How would you categorise your own beliefs about mathematics? Which of Ernest's categories is, in your opinion, most applicable in early childhood mathematics education?

Research highlights the role of teachers' beliefs about mathematics in shaping their mathematics teaching practices (Grootenboer & Marshman, 2016). These connections, as summarised by Beswick (2005; cited in Siemon et al., 2021), are presented in Table 2.2.

TABLE 2.2 Connections between beliefs and mathematics education practices

BELIEFS ABOUT THE NATURE OF MATHEMATICS (ERNEST, 1989)	BELIEFS ABOUT MATHEMATICS TEACHING (VAN ZOEST, JONES, & THORNTON, 1994)	BELIEFS ABOUT MATHEMATICAL LEARNING (ERNEST, 1989)
Instrumentalist	Content-focused, with an emphasis on performance	Skill mastery, passive reception of knowledge
Platonist	Content-focused, with an emphasis on understanding	Active construction of understanding
Problem-solving	Learner-focused	Autonomous exploration of own interests

(Source: Beswick 2005; cited in Siemon et al., 2021)

Just as teachers' beliefs shape their practices, teachers' mathematics education practices in turn shape the mathematical beliefs of the children they are teaching. As Siemon et al. (2021, p. 35) explain, 'even if you are not consciously aware of them, your views about the nature of mathematics will have a powerful influence on how you teach, and hence on how your students learn mathematics and come to view the discipline for themselves'. As such, in becoming a mathematics educator, we may find that a change in our own beliefs about mathematics is required. However, changing beliefs is challenging, and 'usually requires revisiting and reviewing episodes which gave rise to the held beliefs, and then creating new encounters where new and desirable beliefs can be experienced in positive and successful ways' (Grootenboer & Marshman, 2016, p. 17).

ATTITUDES

Attitudes
are learnt responses towards mathematics.

Attitudes can be seen as learnt responses to a situation or object—either positive or negative (Grootenboer & Marshman, 2016). Research has shown that many people appear to have negative attitudes towards mathematics, and these attitudes have largely been developed in school mathematics education (Grootenboer & Marshman, 2016). Ma (1997) identified a reciprocal relationship between attitude and achievement in mathematics and suggested that this relationship results in a self-perpetuating cycle whereby positive attitudes support achievement and negative attitudes impede achievement. The subject of negative attitudes towards mathematics has been studied at both primary and secondary schooling levels, and with pre-service teachers. Bailey (2014) has found that open-ended investigations are effective in improving the mathematical attitudes of pre-service teachers who had initially struggled with their mathematics education courses.

EMOTIONS

Emotions are generally considered to be 'affective responses to a particular situation that are temporary and unstable' (Grootenboer & Marshman, 2016, p. 20). An individual's affective responses may manifest as either positive or negative emotions, some of which are presented in Table 2.3.

Emotions are affective responses towards mathematics.

TABLE 2.3 Positive and negative emotions

POSITIVE EMOTIONS	NEGATIVE EMOTIONS
Enjoyment	Boredom
Anticipatory joy	Hopelessness
Hope	Anxiety
Joy about success	Sadness
Satisfaction	Disappointment
Pride	Shame and guilt
Gratitude	Anger
Empathy	Jealousy and envy
Admiration	Contempt
Sympathy and love	Antipathy and hate
Relief	

(Source: Murphy, MacDonald, Wang, & Danaia, 2019, adapted from Pekrun et al., 2002, p. 92)

Mathematics generates a variety of emotional responses among individuals. These:

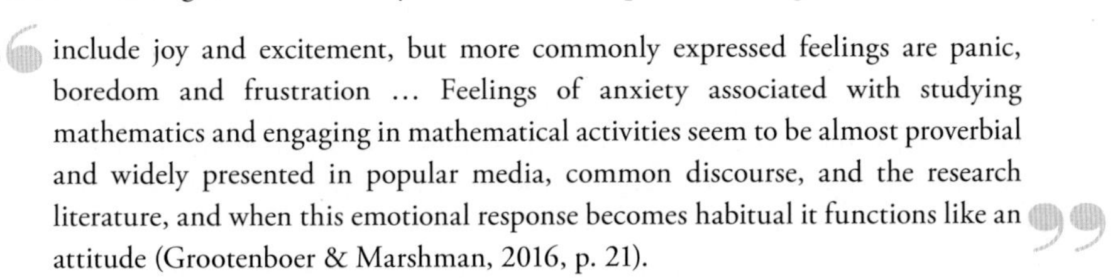

> include joy and excitement, but more commonly expressed feelings are panic, boredom and frustration … Feelings of anxiety associated with studying mathematics and engaging in mathematical activities seem to be almost proverbial and widely presented in popular media, common discourse, and the research literature, and when this emotional response becomes habitual it functions like an attitude (Grootenboer & Marshman, 2016, p. 21).

As indicated in the quote above, anxiety is a common response to mathematics activities. **Mathematics anxiety** is more than just a dislike of mathematics; it refers to feelings of tension and fear that interfere with solving mathematical situations in everyday life (Wilson & Raven, 2014). Mathematics anxiety is common among pre-service teachers, and a great deal of research has been undertaken to ascertain how educators who experience mathematics anxiety can best be supported. Wilson and Raven (2014) have found that bibliotherapy—guided reading of carefully chosen texts—can assist educators in understanding and addressing their mathematics anxiety. Nicolaou, Evagorou, and Lymbouridou (2015) have shown that a range of activities and success in knowledge acquisition helps to activate positive emotions, and Simon et al. (2015) have demonstrated that attaining positive emotions such as enjoyment can help improve persistence and confidence.

Mathematics anxiety is feelings of tension and fear towards mathematical situations.

Have you experienced mathematical anxiety at any stage in your schooling or adult life? What strategies have you used to help manage these feelings of anxiety?

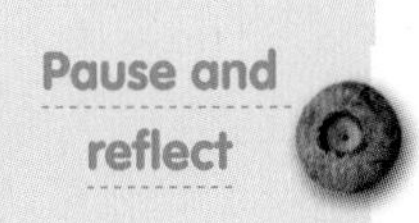

MATHEMATICAL IDENTITY

Mathematics education is fundamentally about developing mathematical knowledge and skills *along with* beliefs, values, attitudes, and feelings about mathematics (Grootenboer & Marshman, 2016):

> Mathematics education facilitates students' development in all these realms, and the nature of education means that this occurs simultaneously. In other words, in mathematics classrooms students are simultaneously learning and developing knowledge, skills and affective responses to mathematics, and this occurs in an integrated manner (p. 27).

Mathematical identity is how people label and understand themselves in relation to mathematics.

Dispositions are the characteristics that encourage children to respond in particular ways to learning opportunities.

Grootenboer and Marshman (2016) use the term **mathematical identity** as a unifying idea to bring together the dimensions of beliefs, values, attitudes, emotions, **dispositions**, cognition, abilities, skills, and life histories. In short, the concept of mathematical identity can be taken to mean how people label and understand themselves (for example, 'I am good at maths'), as well as how an individual is recognised and viewed by others (for example, 'She is good at maths') (Grootenboer & Marshman, 2016).

Memories of learning mathematics

Our own memories of learning mathematics have a profound impact on our mathematical identities. While it is certainly the case that many early childhood educators have fond memories of their own mathematics education, it is well documented that there are large numbers of early childhood educators who experience fear and anxiety in relation to mathematics (Macmillan, 2009), often as a result of their experiences with school mathematics education. Research has found that some educators choose early childhood education rather than primary or secondary teaching because they believe it will not involve teaching mathematics (Macmillan, 2009). However, research has also found that the 'fears and insecurities experienced by many teachers about mathematics dissipate when they share in children's curiosities and engagements with mathematical ideas and processes' (Macmillan, 2009, p. 113).

What are your own memories of learning mathematics? Are these memories predominantly positive or negative? How have these memories shaped your feelings about mathematics, and your feelings about yourself as a mathematics educator?

In Sandra's reflection, which follows, she shares her memories of mathematics—the good and the bad. In particular, her points about feeling silly if you didn't know how to do something, and being embarrassed when asking questions, are important things to keep in mind when working with children and families. Indeed, many children experience these feelings, and so do their parents. Reflecting on your own experiences of mathematics enables you to approach mathematics with children and their families in empathetic and sensitive ways.

Educator reflection

I remember dreading mathematics as it always seemed to be so difficult to learn, and with parents who spoke very limited English and also had very little schooling, I felt that I had no hope. I guess that having a negative attitude to learning maths never made me any better at it. Now as an adult there are still certain types of maths that make me cringe—for example, when trying to help my sister-in-law, who is in Year 9, work out trigonometry, geometry, and algebra. Such situations remind me of just how much I disliked mathematics. I never felt confident when I was a student. I felt silly that I did not know how to do maths and at times I asked questions; however, I would still not understand it and then I was too embarrassed to ask the teacher again.

I have a fond memory of being in Portugal when I was about four years old, helping my grandma get the bananas from her banana trees and then counting how many bananas we had collected. We then split the bananas between ourselves and the neighbours. I also recall the excitement of paying for things and counting money/change from the shops. Not only did it make me feel confident about knowing how to count on my own, but it made me feel independent and capable.

(Source: Sandra Dos Reis)

Learn more
Go online to hear this Educator reflection being read.

For many people, a dislike of mathematics can be linked to a lack of understanding of how mathematics is embedded in our everyday lives. However, learning to recognise mathematical concepts and processes in daily life can help to bring about a new appreciation of mathematics. In Keira's reflection, her comment 'if someone had made that link for me' is particularly profound, especially in the context of this book with its emphasis on noticing and talking about mathematics. Just think what the impact might have been if someone had 'named' an everyday activity as mathematics!

Educator reflection

As a child I remember not liking maths at all! Funny, I remember my mum not liking maths also ... Maths to me was dull; it always had to be written and was difficult. This was the only subject I ever really had trouble with, so obviously I wasn't confident at it. However, I also have a memory of when I was in Grade 2, so I would have been seven years old. My teacher during a whole group time had different-sized containers and was doing lots of pouring and estimating with the group. I can remember absolute excitement over this, a happy memory that has stuck with me. Although at the time I wouldn't have known it was maths. If someone had made that link for me perhaps I would have come to see the joy of maths rather than the dullness.

(Source: Keira Hood)

Learn more
Go online to hear Keira Hood read this Educator reflection.

Indeed, recognising the mathematics embedded in everyday life can have a positive influence on our beliefs about, and attitudes towards, mathematics. Parents can be powerful role models for how mathematics is an essential skill in everyday family life. As Sharon's reflection demonstrates, parents can model positive attitudes towards mathematics and demonstrate the usefulness of mathematics to their children. When modelled to children from a young age, this can help children to gain and sustain an interest in, and appreciation of, mathematics.

Learn more
Go online to hear this Educator reflection being read.

Educator reflection

As a child I liked maths. Maybe because I watched Mum work figures out to keep the household running. I would sit with her and marvel at how she could add this and take away that and juggle the finances to make ends meet. After this she would place cash portions in each bill envelope, rubber band all the envelopes together and write me a list of the places I needed to go to and pay the bills. I was only ten years old, but I loved being able to do my part. Mum would write on each envelope how much change I would get back and I diligently checked for accuracy. Mum could quickly add figures without a calculator, a skill that was passed on to me. Reflecting on Mum exposing me to this early interest and saturation in maths, I can plainly see that by introducing mathematical concepts to young children, the way is made much easier and more natural.

(Source: Sharon Cope)

A key to sustaining children's interest in mathematics as they progress through their formal education is making the mathematics *meaningful* and relevant to daily life. Many of us will recall our own experiences of formal mathematics education and the situations in which we found ourselves wondering, 'Why do I need to know this?'

Learn more
Go online to hear Aimie Gibson read this Educator reflection.

Educator reflection

I liked maths in school; I was able to memorise the times tables easily and always felt confident. In high school I did advanced maths and then extension, but found the extension maths didn't really have any contextual reference to my day-to-day life, and ended up being stressful so I dropped it after a year. I do more basic everyday maths these days rather than in-depth calculations, but I definitely still have a positive disposition towards teaching children maths that is relevant to them and can help them feel knowledgeable and empowered—for example, telling the time or being able to move their own counter when playing board games. I'd like to think this positive disposition and knowledge of mathematical concepts in daily practices have helped my boys learn maths. I thought my 6.5-year-old son was great at maths but my 4.5-year-old loves numbers and is even more proficient as he has learnt from his older brother.

(Source: Aimie Gibson)

In this reflection, Aimie makes a critical point about contextual relevance. For many, a lack of relevance (or meaningfulness) serves as the switching-off point. Aimie also makes a powerful point about the importance of dispositions, and she clearly articulates how her interest and positive attitude have had a great impact on her children's engagement with mathematics. This is a key message we can deliver through our mathematics education practices. While not all of us have had great experiences with mathematics, and some may not feel confident with it, if we can display interest in and react with positivity towards *children's* engagements with mathematics, we can help them to have different experiences to those negative ones that may have plagued us. And perhaps we can change our own perceptions in the process.

Mathematical mindsets

A commonly held belief is that mathematics is something that you are 'just good at' (or not good at). You are likely to have heard someone make a comment about 'having a maths brain'—you may even have made such a comment yourself. Such comments reflect a belief that achievement in mathematics depends mostly on *aptitude* or *ability* (Clements & Sarama, 2014). On the other hand, it is possible to think of mathematics achievement as coming from *effort*—that is, the belief that if you try hard and work hard you can achieve in mathematics (Clements & Sarama, 2014).

People who believe that you either 'get it or you don't' often develop something known as 'learnt helplessness' (Clements & Sarama, 2014). This can result in a self-perpetuating cycle of difficulty in mathematics education because the individual believes that they will *always* struggle with mathematics. However, even those with learnt helplessness in mathematics can be supported to see themselves as capable of achieving in mathematics. Dweck (2006) proposed the notion of a **growth mindset**—the belief that intelligence and 'smartness' can be learnt and that the brain can grow from practice and challenge. As Boaler (2013) explains:

Growth mindset is the belief that intelligence can be learnt.

> People with a growth mindset work and learn more effectively, displaying a desire for challenge and resilience in the face of failure. On the other hand, those with a 'fixed mindset' believe that you are either smart or you are not. When students with a fixed mindset fail or make a mistake they believe that they are just not smart and give up (p. 143).

The impact of fixed mindsets is so often seen in mathematics education, with people believing they 'are just not good at maths' and giving up. As such, researcher Jo Boaler has conducted extensive work around growth mindsets in mathematics—what she describes as 'mathematical mindsets'. **Mathematical mindset** expresses the idea that we are all capable of learning and improving in mathematics. Mistakes are not seen as failures, but rather as opportunities to learn. Boaler (2013) suggests that the tasks teachers choose allow different messages to be communicated to students:

Mathematical mindset is the belief in our ability to learn and improve in mathematics.

> if students are working on short, closed questions that have right or wrong answers, and they are frequently getting wrong answers, it is hard to maintain a view that high achievement is possible with effort. When tasks are more open, offering opportunities for learning, students can see the possibility of higher achievement and respond to these opportunities to improve (p. 146).

The notion of mathematical mindsets provides one possible explanation for why we think and feel about mathematics the ways that we do. The ways we have experienced mathematics as a student will contribute to our having either a *fixed mindset* or a *growth mindset* about our own mathematics ability. Self-awareness can help us to better understand how the children with whom we work respond to the mathematical learning experiences with which they are faced. Careful attention to the design of these experiences can support children to see mathematics as something that they can practice and improve at, and to appreciate its application to everyday life.

Learn more
Go online to hear Alexandra Roth read this Educator reflection.

Educator reflection

I have never liked or appreciated mathematics. I don't think I have a maths brain but I am not scared by it. At times I'm awed by it, particularly the Fibonacci sequence in nature. I relate best to maths when it has a practical application; for instance, when I go shopping and compare prices.

(Source: Alexandra Roth)

Chapter summary

This chapter has encouraged you to reflect on the affective domain of mathematics and consider your own memories of mathematics and your mathematical identity. An understanding of our own mathematical identity is important as this influences our approaches as mathematics educators, which in turn influence the children and families with whom we work. This chapter has advocated for the importance of 'mathematical mindsets' that help us to feel that we can always learn, and improve, in mathematics. Having a positive disposition towards mathematics helps us, as educators, to encourage children to have positive dispositions towards mathematics.

The specific topics we covered in this chapter were:

- the affective domain of mathematics education
- beliefs
- attitudes
- emotions
- mathematical identity
- memories of learning mathematics
- mathematical mindsets.

FOR FURTHER REFLECTION

1. Can you recall a time when you felt excited by, and confident with, mathematics?
2. What were the factors that enabled you to feel confident as a mathematician?
3. How can you apply this to your own mathematics teaching practices to help the children with whom you work feel confident and excited?

FURTHER READING

Attard, C., Ingram, N., Forgasz, H., Leder, G., & Grootenboer, P. (2016). Mathematics education and the affective domain. In K. Makar, S. Dole, J. Visnovska, M. Goos, A. Bennison, & K. Fry (eds), *Research in mathematics education in Australasia 2012–2015* (pp. 73–96). Singapore: Springer.

Boaler, J. (2013). Ability and mathematics: The mindset revolution that is reshaping education. *Forum, 55*(1), 143–52.

Grootenboer, P., & Marshman, M. (2016). The affective domain, mathematics, and mathematics education. In *Mathematics, affect and learning* (pp. 13–33). Singapore: Springer.

03 Theoretical Perspectives

Learn more
Go online to see a video from Paige Lee explaining the key messages of this chapter.

CHAPTER OVERVIEW

This chapter will explore a number of theoretical perspectives that can inform early childhood mathematics education. These theories explore different aspects of children's development that provide bases for, and insights into, children's mathematical learning. A general explanation of each perspective will be provided, along with discussion of how each perspective is relevant to, and useful for, approaches to mathematics education in early childhood settings.

In this chapter, you will learn about the following topics:

- developmental theories
- constructivism
- social constructivism
- ecological theory
- representation theory
- constructionism
- self-regulation.

KEY TERMS

- Bi-directional influences
- Centration
- Chronosystem
- Classification
- Cognition
- Concrete operational period
- Conservation
- Constructionism
- Constructivism
- Culture
- Developmental theory
- Ecological theory
- Enactive representations
- Exosystem
- External representations
- Formal operational period
- Iconic representations
- Internal representations
- Irreversible thinking
- Logico-mathematical knowledge
- Loose parts
- Macrosystem
- Mesosystem
- Microsystem
- Perceptual dominance
- Physical knowledge
- Preoperational period
- Representation theory
- Representations
- Scaffolding
- Self-regulation
- Sensorimotor period
- Seriation
- Social constructivism
- Social-conventional knowledge
- Symbolic representations
- Symbolism
- Whole child
- Zone of Proximal Development (ZPD)

Introduction

This chapter presents a selection of theoretical perspectives that can inform early childhood mathematics education. Theoretical perspectives can help us to make sense of the many ways in which young children develop mathematical knowledge, and use that knowledge in different ways in a variety of contexts. The theories discussed in this chapter present varying views of children's learning and development. Some focus on stages of development, while others emphasise social or socio-cultural perspectives of children's learning. While a holistic view of children's development is of the upmost importance in early childhood education, children's *cognitive* development is of particular relevance to early childhood mathematics education. **Cognition** refers to mental processes, such as reasoning, planning, problem solving, representing, and remembering (Arthur et al., 2012). The theoretical perspectives presented in this chapter attend largely to children's cognitive development; though, other aspects of development—namely, social, language, emotional—are certainly reflected within these perspectives.

Cognition refers to mental processes, such as reasoning, planning, problem solving, representing, and remembering.

The importance of using different theories to inform our understandings of children's learning and development is recognised in the *Early Years Learning Framework for Australia* (EYLF; DET, 2019, p. 12). It states:

> Early childhood educators draw upon a range of perspectives in their work which may include:
>
> - developmental theories that focus on describing and understanding the processes of change in children's learning and development over time
> - socio-cultural theories that emphasise the central role that families and cultural groups play in children's learning and the importance of respectful relationships and provide insight into social and cultural contexts of learning and development
> - socio-behaviourist theories that focus on the role of experiences in shaping children's behaviour
> - critical theories that invite early childhood educators to challenge assumptions about curriculum, and consider how their decisions may affect children differently
> - post-structuralist theories that offer insights into issues of power, equity and social justice in early childhood settings.
>
> Drawing on a range of perspectives and theories can challenge traditional ways of seeing children, teaching and learning, and encourage educators, as individuals and with colleagues, to:
>
> - investigate why they act in the ways that they do
> - discuss and debate theories to identify strengths and limitations
> - recognise how the theories and beliefs that they use to make sense of their work enable but also limit their actions and thoughts
> - consider the consequences of their actions for children's experiences
> - find new ways of working fairly and justly (p. 12).

It is important to recognise that adequate explanations of children's lives and experiences often involve more than one theoretical approach (Arthur et al., 2015). While there are certainly strengths and weaknesses associated with the differing perspectives presented in this chapter, all offer valuable insights into the different ways in which children develop and experience mathematical understandings in their early years.

Developmental theories

Early childhood educators draw on a range of theories of development in their work with young children, and these many theories contribute to the knowledge base of early childhood education (Arthur et al., 2012). **Developmental theories**, in general, focus on stages of development that are sequential and somewhat predictable in nature and are characterised by specific developmental tasks and accomplishments (Kearns, 2010). Development is typically considered in relation to five separate developmental domains: physical, social, emotional, language, and cognitive. Kearns (2010, p. 26) offers the following summary of the key concepts underpinning developmental theories:

Developmental theories focus on sequential, predictable stages of development.

- Development occurs in stages that are generally age related. Each stage is defined by a set of 'typical' skills or understandings that children would normally be expected to achieve at or around a particular age.
- Development at one stage lays the base for later development. Children must be allowed time for learning at each stage.
- Development moves from simple to complex and from general to specific. Children's development becomes more complex as they move through each stage of development.
- Development occurs in a predictable sequence.
- Individuals develop according to a particular timetable and pace. The timing and length of each stage can vary from one individual to another; however, the sequence remains the same.
- The terms 'normal', 'age-appropriate', or 'typical' development are generally used to describe development that falls within the 'usual' time range for the development of skills and behaviours.
- There are optimal periods in development. Learning occurs most easily when children are developmentally 'ready'. If a child is not 'ready', then no amount of 'teaching' will make learning happen easily.
- Development results from the interaction of biological factors (maturation) and environmental factors (learning). As the child matures, he or she will become 'ready' to learn new things. The environment or circumstances in which the child lives will affect his or her opportunities to learn.
- One area of development affects and influences another area of development.

Developmental theory is evident in approaches to mathematics education that focus on 'ages and stages'—that is, what children 'can' or 'should' learn at particular ages. This type of thinking is reflected in many curriculum documents, particularly in the schooling years, and is evident in planning documents such as scope and sequence charts for mathematics. While there is certainly value in having a sense of sequence and trajectory for children's mathematical learning, the risk is that children's learning opportunities may be limited based on a view of what they can and cannot do at particular ages. Recent research has demonstrated that curriculum documents often underestimate children's mathematical capacities, particularly in relation to children's mathematical knowledge upon school entry (Gervasoni & Perry, 2015). Indeed, research has shown that many children commence primary school already knowing what the curriculum identifies should be *taught* to them in their first year of school (Gervasoni & Perry, 2015; Gould, 2012). In short, it is important to consider what mathematical ideas might be *accessible* to children at different ages, but not limit the range of possibilities for individual children on the basis of preconceived ideas of what mathematics children *can* or *should* access.

What are your personal feelings about the role of developmental theories in understanding children's mathematical learning?

Constructivism

Constructivism
is based on the belief that learners construct their own knowledge.

Social-conventional knowledge
is information gained through direct social transmission.

Physical knowledge
is information about the qualities or attributes of objects and what they are used for.

Logico-mathematical knowledge
is developed in the mind by thinking about an object.

Sensorimotor period
involves understanding the present and the real world.

Preoperational period
involves symbolic representation of the present and real.

Concrete operational period
involves the organisation of concrete operations.

Formal operational period
involves hypothesis making and testing the possible.

Jean Piaget (1896–1980) is perhaps the most significant theorist in relation to children's cognitive development, and his work has certainly had a direct connection to, and lasting impact upon, mathematics education through its focus on the development of thinking, reasoning, memory, and logic. Piaget argued that children's thinking reflects their own unique way of understanding and interpreting the world (Arthur et al., 2012). Underpinning Piaget's theoretical work is the belief that learners construct or develop their own knowledge—a perspective known as **constructivism**. Piaget's theory is based on the belief that children do not think in the same way as adults, and that children use different cognitive processes and skills to construct three types of knowledge:

- **social-conventional knowledge**—information gained through direct social transmission
- **physical knowledge**—information about the qualities or attributes of objects and what they are used for; physical knowledge is acquired by sensory exploration, learnt facts, experimenting, and observing
- **logico-mathematical knowledge**—knowledge that is developed in the mind by thinking about an object (as opposed to knowledge that is socially constructed). This knowledge enables children to develop ideas about relationships between objects, and is an essential skill for reasoning (Kearns, 2010, p. 28).

Piaget's constructivist theory of cognitive development focuses on how knowledge is acquired and the particular order in which different ways of thinking develop (Kearns, 2010). Piaget believed that children's development occurs in four predictable, sequential stages, and that all children pass through the same stages in the same order at roughly the same age (Kearns, 2010). Piaget's four stages of development are as follows:

- **sensorimotor period** (0–2 years of age)—understanding the present and the real world
- **preoperational period** (2–7 years of age)—symbolic representation of the present and real; preparation for understanding concrete operations
- **concrete operational period** (7–12 years of age)—organisation of concrete operations; the development of many mathematical concepts
- **formal operational period** (12–15 years of age)—hypothesis making; testing the possible (Kearns, 2010).

Piaget's notion of the preoperational period is of particular interest to early childhood mathematics education as it focuses on the development of cognitive skills that lay the foundation for abstract, logical thought (Kearns, 2010). The preoperational period is characterised by symbolic thought, concept acquisition, reasoning, and information processing. Preoperational skills that relate specifically to mathematics education are outlined in Table 3.1.

TABLE 3.1 Preoperational skills relevant to mathematics education

CHARACTERISTIC	SKILLS
SYMBOLIC THOUGHT	**Symbolism**—the ability to represent things with symbols, including the use of language
CONCEPT ACQUISITION	**Seriation**—the ability to sequence based on specific attributes **Classification**—the process of grouping objects according to specific attributes **Conservation**—the ability to understand that attributes such as weight, length, and number remain constant despite changes in appearance
REASONING	Reasoning from particular to particular Presuming causal relationships if events are associated closely in time or some other way
INFORMATION PROCESSING	**Irreversible thinking**—the inability to begin at the end of an operation and work back to the start **Perceptual dominance**—the tendency to focus attention on, and to judge on the basis of, visually striking features Short-term and long-term memory **Centration**—the tendency to fixate on a single attribute or feature to the exclusion of others

(Adapted from Kearns, 2010, pp. 36–7)

Symbolism is the ability to represent things with symbols, including language.

Seriation is the ability to sequence based on specific attributes.

Classification is the process of grouping objects according to specific attributes.

Conservation is the understanding that attributes such as weight, length, and number remain constant despite changes in appearance.

What examples of these skills have you observed in your interactions with your children?

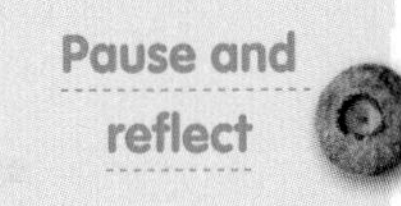

It is well documented that there has been some reluctance among early childhood educators to embrace mathematics within their educational programs (Sarama & Clements, 2002; Perry & Dockett, 2008). This reluctance can, in part, be explained by the predominance of Piagetian perspectives (Arthur et al., 2012). However, more recent research in relation to children's mathematical development has articulated a shift in thinking, with young children now celebrated as capable mathematical thinkers and learners (MacDonald, 2015b). Indeed, a contemporary criticism of Piaget's theory is that many children demonstrate skills associated with particular developmental periods before the age range specified by Piaget. For example, Figure 3.1 shows a fourteen-month-old child demonstrating the preoperational skill of classification during meal time, sorting his meal into groups (i.e. Cheerios, cheese, toast). According to Piaget's theory, such preoperational knowledge would not 'normally' be expected until 2–7 years of age—however, many would argue that the kind of skill depicted in Figure 3.1 is indeed very 'normal' behaviour for a child of this age.

Irreversible thinking is the inability to begin at the end and work backwards.

Perceptual dominance is the tendency to focus on visually striking features.

Centration is the tendency to fixate on a single attribute.

FIGURE 3.1 An infant demonstrating classification skills at meal time

(Source: Amy MacDonald)

These sorts of criticism led to some discontent with Piagetian theory; in particular, with its emphasis on individual development and universal stages. The following are now known:

- Young children are more competent than as described by Piaget.
- Children's cognitive development is not nearly as stage-like as described by Piaget.
- Children's prior knowledge and experiences influence their thinking and reasoning.
- Culture influences cognitive development (Arthur et al., 2015).

In response to this, alternative theoretical perspectives emphasising children as unique, social beings who are influenced by their social and cultural worlds have been embraced by many.

Social constructivism

Russian psychologist Lev Vygotsky (1896–1934) was a contemporary of Piaget and is a commonly cited theorist in mathematics education. In contrast to Piaget's theory of universal stages, Vygotsky believed that cognitive development was a process of internalising ideas that are experienced as a result of interactions with the social and cultural world (Arthur et al., 2012; Kearns, 2010). This socio-cultural theoretical perspective is typically referred to as **social constructivism**. Vygotsky saw learning and development working together as a dynamic process in a socio-cultural and historical context that operated on three levels: 1. the immediate interactive level; 2. the structural level; and 3. the more general cultural or social level. Vygotsky's three levels can be summarised in the following detail:

Social constructivism is based on the belief that cognitive development results from interactions with the social and cultural world.

- The *immediate interactive level*, which embraces two ways in which we develop cognitively through involvement in our social context. The first refers to the construction of our understanding through our interaction with others; and the second to the other more solitary interactions that we have with artefacts and materials in the learning context.
- The *structural level*, which includes social structures that influence the child, such as their family and their early childhood service or school. Within this level, the child encounters new ideas in formal and informal contexts, which reflect the beliefs and values of the family, the educational setting, and the wider community.

- The *more general cultural level*, which represents our social, cultural, and historical interactions. The learner reflects the culture in which he or she is situated (Brooks, 2004).

Vygotsky (1978) argued that all learning is essentially social in origin and that, as well as learning from their teachers, children learn from their families, peers, and adults generally. Unlike Piaget, Vygotsky believed that more knowledgeable others play a crucial role in the development of higher-order knowledge and skills, stressing the importance of social interactions, role modelling, and tutoring for cognitive development (Kearns, 2010). Vygotsky used the term **Zone of Proximal Development (ZPD)** to describe the difference between what a child can learn without assistance and the learning that could be achieved with support from a more knowledgeable person (be it a parent, teacher, or peer) (Kearns, 2010). As shown in Figure 3.2, the ZPD represents *the learning that is possible* with support and scaffolding. Though not explicitly used by Vygotsky himself, **scaffolding** is the term used to describe the process whereby a more knowledgeable other builds on the child's existing knowledge and skills to introduce new or more complex concepts, introduce new language, and challenge the child's thinking (Kearns, 2010). Scaffolding provides the framework or support that enables children to try out new ideas (Arthur et al., 2012), and adults can scaffold children's thinking about mathematical concepts through their everyday activities and interactions (for example, Figure 3.3).

Zone of Proximal Development (ZPD) is the difference between what a child can learn with and without assistance.

Scaffolding is a process of building on a child's existing knowledge to introduce more complex knowledge.

Can you think of an occasion when a teacher scaffolded your own mathematical learning? How was this scaffolding provided?

Pause and reflect

FIGURE 3.2 Zone of Proximal Development (ZPD)

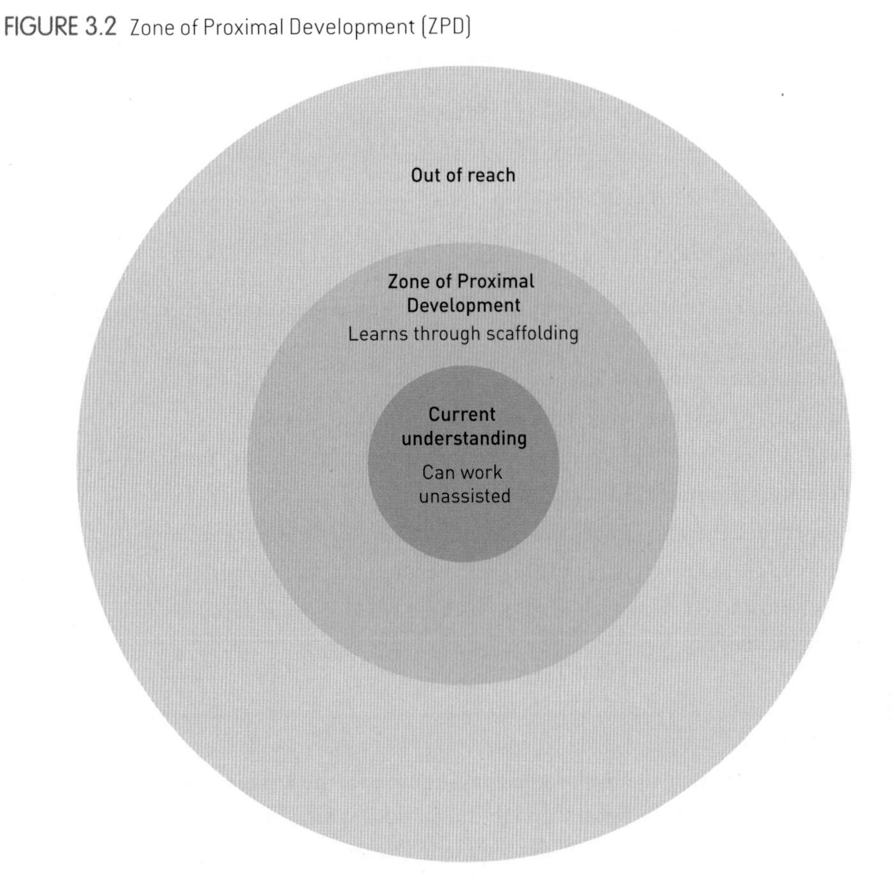

FIGURE 3.3 Scaffolding to support development of mathematical concepts

(Source: Paige Lee)

CURRICULUM CONNECTIONS

Early Years Learning Framework

Outcome 1: Children have a strong sense of identity

Children develop their emerging autonomy, inter-dependence, resilience and sense of agency.

This is evident when children:

- Are open to new challenges and discoveries
- Persist when faced with challenges and when first attempts are not successful.

Educators promote this learning when they:

- Display delight, encouragement and enthusiasm for children's attempts
- Support children's efforts, assisting and encouraging as appropriate
- Motivate and encourage children to succeed when they are faced with challenges.

(DET, 2019, p. 25)

Vygotsky emphasised the importance of language in the learning process and argued that language is made meaningful through interactions with adults or peers who can assist children to reach a higher level of development (Kearns, 2010). Vygotsky described all higher mental functions as having social origins—that is, they appear first in interactions between people before they are internalised by individuals (Arthur et al., 2012). Vygotsky emphasised the social and cultural contexts of language development and believed that thought was the result of language—unlike Piaget, who believed that cognitive development determined language use (Kearns, 2010).

Building on Vygotsky's social constructivist theory, Rogoff (1990) has emphasised a socio-cultural view of learning and development. Rogoff argues that all children's learning reflects those things that are important in the child's culture. Rogoff (2003) defines **culture** as the common ways of knowing and being that people in a community share. Children's learning is influenced by the cultural patterns of the communities in which they participate, and the interactions and discourses within those communities. Socio-cultural views of development reflect a notion of the **whole child**—that is, a recognition that each child is a unique individual who is shaped by many influences (Kearns, 2010).

Culture
can be defined as the common ways of knowing and being that people in a community share.

Whole child
refers to the recognition that each child is a unique individual who is shaped by many influences.

Ecological theory

A theorist whose work encompasses the notion of the whole child is Urie Bronfenbrenner (1917–2005), whose **ecological theory** of development is based on the idea that children's development is influenced by various social and cultural systems within the environment, or 'ecology', in which they live. Bronfenbrenner's ecological theory shares many features with Vygotsky's social constructivist view of learning and development. Like the social constructivist perspective, ecological theory grew out of a discontent with the prevailing models of development, in which children's development was being studied 'out of context' (Bronfenbrenner, 1988). To combat this, Bronfenbrenner designed a model of development that focuses on the contexts of children—such as their home, educational setting, and community—and the socially and culturally based connections that prevail between these contexts and that may endure or fluctuate over time.

Ecological theory
is based on the belief that children's development is influenced by the social and cultural systems within the child's environment.

As shown in Figure 3.4, Bronfenbrenner's theory positions the child at the centre of the model and acknowledges that while biological and environmental factors affect a child's development, each child actively contributes to their own development as a result of the interactions and relationships they have within five different systems (Kearns, 2010).

The **microsystem** is made up of the immediate environments in which the child participates—for example, their home, early childhood service or school, local neighbourhood, and so forth. The microsystem also consists of the roles and interpersonal relationships within these settings. Bronfenbrenner used the term **bi-directional influences** to describe the relationships between the child and the various settings in which the child participates (Kearns, 2010). These relationships both influence and are influenced by the child.

Microsystem
is made up of the immediate environments in which the child participates.

Bi-directional influences
are the relationships between the child and the settings in which the child participates.

The **mesosystem** involves relationships between microsystems, or connections between contexts. An important mesosystem for young children is the relationship between their home and their educational setting (early childhood service or school). According to Bronfenbrenner (1979), we can expect more enhanced development when this mesosystem is characterised by more frequent interaction between parents and teachers, a greater number of persons known in common by members of the two settings, more frequent communication between home and educational setting, and more information in each setting about the other.

Mesosystem
involves relationships between microsystems.

FIGURE 3.4 Bronfenbrenner's ecological theory

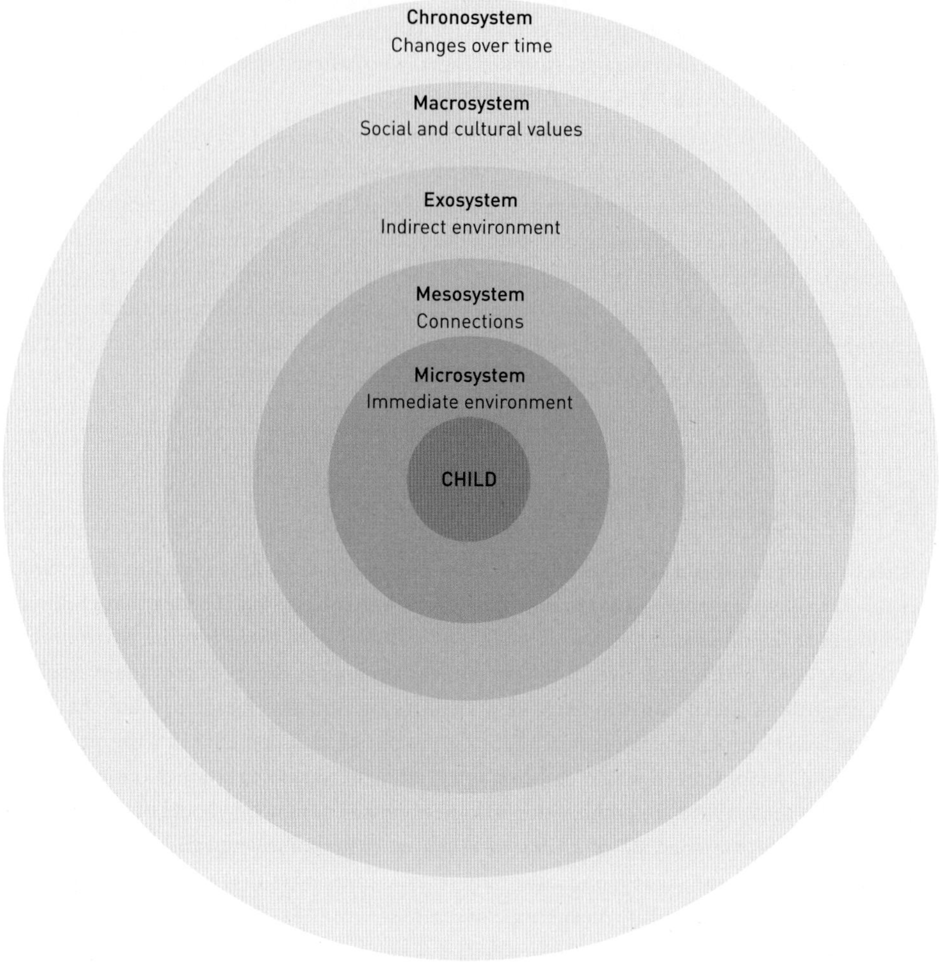

Chronosystem
Changes over time

Macrosystem
Social and cultural values

Exosystem
Indirect environment

Mesosystem
Connections

Microsystem
Immediate environment

CHILD

Exosystem
is the social system one step removed from the child.

The **exosystem** is the child's indirect environment—the social system one step removed from the child (Kearns, 2010). The child's exosystems are those that have power over their life, yet in which they do not participate. For example, a parent's employment, and access to family and community services, have an indirect impact upon the child's development because of their connection with the child's family microsystem (Kearns, 2010). Similarly, education systems, curriculum directorates, and the like impact the child's development because they make decisions that affect the child's education microsystem (i.e. early childhood service or school).

Macrosystem
is the wider cultural, social, and political context.

Microsystems, mesosystems, and exosystems are embedded within the **macrosystem**—the wider cultural, social, and political context. The macrosystem influences the child through the cultural values, laws, and customs of the community in which the child lives (Kearns, 2010). The macrosystem establishes the norms about how development proceeds and the appropriate nature and structure of microsystems, mesosystems, and exosystems (Garbarino & Plantz, 1980).

Chronosystem
is the socio-historical timeframe of the child's life.

Finally, the **chronosystem** is the social and historical timeframe in which the child's life is set (Kearns, 2010). This system involves the patterning of environmental effects and transitions over the life course, as well as socio-historical circumstances. The chronosystem reflects how children change over time (Kearns, 2010).

Bronfenbrenner's ecological theory is useful for mathematics education because it encourages us to consider, comprehensively, the socio-cultural spheres of influence that contribute to a child's development of mathematical understandings. By viewing mathematical learning through an ecological lens, educators can consider the *specific* components that interact in order to shape the child's mathematical development. An ecological perspective allows us to see the importance of the people, places, activities, and cultures that serve as the contexts for mathematical learning, and thus come to acknowledge that these contexts endow mathematics with *meaning*.

Think about your own learning of mathematics. Can you identify influences from each of Bronfenbrenner's 'systems'?

CURRICULUM CONNECTIONS

Early Years Learning Framework

Outcome 1: Children have a strong sense of identity

Children develop knowledgeable and confident self identities.

This is evident when children:

- Feel recognised and respected for who they are
- Share aspects of their culture with the other children and educators.

Educators promote this learning when they:

- Show respect for diversity, acknowledging the varying approaches of children, families, communities and cultures
- Acknowledge and understand that children construct meaning in many different ways
- Demonstrate deep understanding of each child, their family and community contexts in planning for children's learning.

(DET, 2019, p. 26)

Representation theory

Representation theory, like social constructivist theory and ecological theory, also draws attention to the role of context in developing mathematical understandings. **Representation theory** focuses on the construction of representations, and the contexts that influence the ways these representations are constructed. In mathematics education, it can be useful to think about **representations** as images, signs, characters, or objects that stand for or symbolise something. Goldin and Shteingold (2001) offer the following example:

Representation theory is focused on the construction of representations and the contexts that influence these representations.

Representations are images, signs, characters, or objects that stand for or symbolise something.

> The numeral 5 can represent a particular set containing five objects, determined by counting; or it can stand for something much more abstract—an equivalence class of such sets. It can also represent a location or the outcome of a measurement …

> So we see that the thing represented can vary according to the context of the use of the representation (p. 3).

Sharing the constructivist view that children are active participants in the construction of their own knowledge, American psychologist Jerome Bruner (1915–2008) emphasised the role of representation in the learning process. Bruner (1966) proposed three modes of representation:

Enactive representations use real materials to depict relationships.

Iconic representations use drawings or pictures to depict relationships.

Symbolic representations use abstract symbols to depict relationships.

- **enactive representation**—where the child can manipulate real materials to depict relationships
- **iconic representation**—where the child can work with drawings or pictures representing relationships
- **symbolic representation**—where abstract symbols are understood to describe relationships.

Vygotsky also emphasised the critical role of representation in young children's concept development and viewed representation as a 'way of knowing'. For young children, representation activities such as drawing help to bring ideas to the surface (Woleck, 2001). Bruner argued that children are actively engaged in 'meaning making' through their interactions with people and their environments (Kearns, 2010). He referred to both *constructing* meaning and the *processing* of information as a way of understanding development (Kearns, 2010).

Internal representations are mental configurations that are not directly observable.

External representations are physical, observable configurations.

Reflecting this notion of constructing and processing, Goldin and Kaput (1996) viewed development as a process of translating between *internal* and *external* representations. **Internal representations** are the mental configurations of individuals that are not directly observable. **External representations** are physical, observable configurations that can be apprehended by anyone with suitable knowledge. Put simply, an internal representation is the mental image held by an individual about a concept as a result of their experiences or observations, while an external representation is an artefact by which this mental image is communicated. For example, a child's drawing can be considered an *external* representation that is used to communicate an *internal* representation or mental image. Drawings are

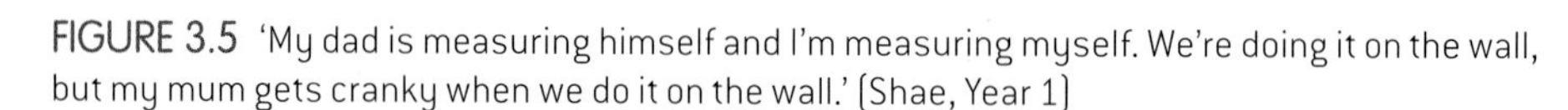

FIGURE 3.5 'My dad is measuring himself and I'm measuring myself. We're doing it on the wall, but my mum gets cranky when we do it on the wall.' (Shae, Year 1)

(Source: Amy MacDonald)

particularly useful for gaining understanding about not only *what* children know about mathematics, but *how* they know it—that is, the experiences that have helped them to develop those understandings. As shown in Figure 3.5, creating a drawing (i.e. an external representation) is a way for a child to share their mental image or memory (i.e. an internal representation) of how they came to know about a particular mathematical idea.

An important consideration is the two-way interactions between internal and external representations (Figure 3.6). As Goldin and Kaput (1996) explain, sometimes an individual externalises knowledge in a physical form as a result of their internal understandings. However, sometimes an individual internalises an understanding as a result of interactions with external, physical materials and experiences. These two-way interactions can take place at both an active, deliberate level, and at a more passive, automatic level—for example, sometimes mathematical language is just 'understood' without deliberate, conscious mental activity. Furthermore, interactions between internal and external representations, in both directions, can (and most often do) occur simultaneously (Goldin & Kaput, 1996).

Consistent with Piaget's constructivist theory, new systems of representation are typically built up from existing systems (Goldin & Shteingold, 2001). Goldin and Kaput (1996) identify three main stages in the development of new representations:

- First, the inventive-semiotic stage operates, during which new characters or symbols are introduced. They are used to symbolise aspects of a previously developed representational system, which is the basis for their 'meaning'.
- Second, the earlier system is used as a template for the structure of the new system. Rules for the new symbols are worked out, using the earlier system together with the new meanings that have been assigned.
- Third, the new system becomes autonomous. It can be detached from the template that helped to produce it and can acquire meanings and interpretations different from, or more general than, those that were first assigned.

FIGURE 3.6 Interactions between internal and external representations

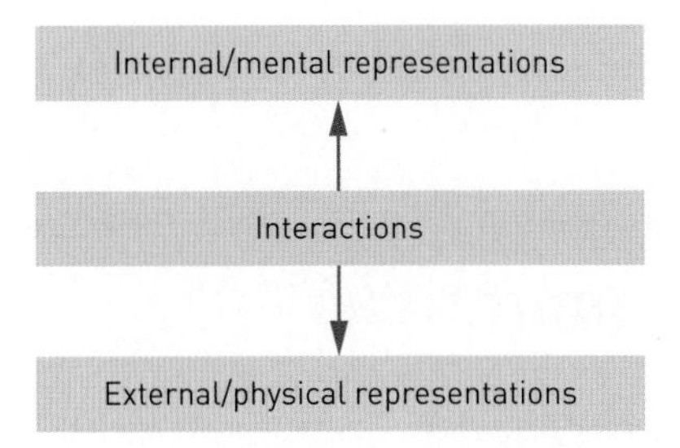

Consider, for example, a child who is learning to use written symbols to solve addition problems. The child may have a previous representational system in which they have used concrete objects to create an external representation of the addition problem, as shown in Figure 3.7. The child can use this representation system as the template for a new system, where numerals and symbols are instead used to represent the addition problem, i.e. $5 + 2 = 7$. Eventually, the child is able to detach this new system from the earlier system—that is, the child can solve the addition problem based on the numerals and symbols alone, without needing to model it with concrete materials.

FIGURE 3.7 Developing new representations

Representations are increasingly being seen as useful tools for both constructing and communicating mathematical understandings, and they play an important role in children's mathematical activity (Siemon et al., 2021). As educators, we need to:

- utilise representations that best support children's learning
- assist children to draw upon a range of representations to explore mathematical concepts
- encourage the use of language to explain and justify mathematical thinking to others
- assist children to discover the mathematics embedded in representations (Siemon et al., 2021).

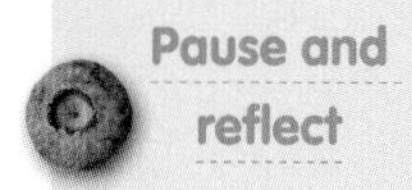

Pause and reflect

What other examples of mathematical representations can you think of?

CURRICULUM CONNECTIONS

Early Years Learning Framework

Outcome 4: Children are confident and involved learners

Children develop a range of skills and processes such as problem solving, inquiry, experimentation, hypothesising, researching and investigating.

This is evident when children:

- Create and use representation to organise, record and communicate mathematical ideas and concepts.

Educators promote this learning when they:

- Encourage children to make their ideas and theories visible to others.

(DET, 2019, p. 38)

Constructionism

Closely related to representation theory is the constructionist perspective. The theory of constructionism (with an *n* rather than a *v*) emerged in the 1980s from the work of Seymour Papert (1928–2016), a protégé of Piaget. Papert's **constructionism** theory builds on the constructivist notion that knowledge is actively constructed in the mind of the learner and extends this to suggest that learners are particularly inclined to generate new ideas when they are actively engaged in making some type of artefact (Brady, 2008)—that is, when they are *constructing* something. As Brady (2008, p. 77) explains, constructionism involves two intertwined types of construction: the construction of knowledge in the context of constructing personally meaningful artefacts (what Papert referred to as 'objects-to-think-with'); and the creation of an artefact that will be viewed and valued by others—an artefact that facilitates reflection and sharing.

Constructionism is based on the belief that learners generate new ideas when they are engaged in constructing an artefact.

Constructionist perspectives promote concrete representations of mathematical ideas. Children are encouraged to make their mathematical thinking visible and tangible through the production of models, diagrams, graphs, drawings, etc. Manipulation of concrete materials is particularly encouraged, and constructionist approaches make use of activities such as paper folding, cutting, and rearranging to model mathematical ideas and processes (see Figure 3.8).

FIGURE 3.8 A constructionist approach

(Source: Amy MacDonald)

The constructionist theoretical perspective is compatible with the theory of 'loose parts', which is popular in early childhood education. **Loose parts** are materials that can be moved, carried, combined, redesigned, lined up, and taken apart and put back together in multiple ways ('Let the children play', 2010). Loose parts are important for construction play. They allow children to make, invent, construct, evaluate, and modify (Maxwell, Mitchell, & Evans, 2008); all of which facilitates mathematical thinking, representation, and knowledge construction.

Loose parts are materials that can be moved, combined, taken apart, and put together in multiple ways.

CURRICULUM CONNECTIONS

Early Years Learning Framework

Outcome 4: Children are confident and involved learners

Children resource their own learning through connecting with people, place, technologies, and natural and processed materials.

This is evident when children:

- Manipulate resources to investigate, take apart, assemble, invent and construct.

Educators promote this learning when they:

- Provide sensory and exploratory experiences with natural and processed materials
- Provide opportunities for children to both construct and take apart materials as a strategy for learning
- Provide resources that encourage children to represent their thinking.

(DET, 2019, p. 40)

Self-regulation

Self-regulation is the capacity to use thought to guide behaviour.

Mathematical thinking and problem solving involve taking in and interpreting information, operating on it, and responding to it (Clements & Sarama, 2014). Attention is key to this process, and the broader competence that helps children to focus their attention is self-regulation (Clements & Sarama, 2014). **Self-regulation** is the capacity to use thought to guide behaviour (Arthur et al., 2012). More specifically, self-regulation is (Clements & Sarama, 2014):

> the process of intentionally controlling one's impulses, attention, and behaviour. It may involve avoiding distractions, and maintaining a focus on setting goals, planning, and monitoring one's attention, actions, and thoughts (p. 238).

Research has begun to identify the role that self-regulation plays in children's learning, generally (for example, Blair, Calkins, & Kopp, 2010); and in relation to the cognitive demands of mathematics, specifically (for example, Bull & Lee, 2014; Ivrendi, 2011). Recent research has shown that children's attentional regulation is directly associated with mathematics achievement from four to five years of age (Williams, White, & MacDonald, 2016). Self-regulation is an important skill for mathematics development because mathematical work 'requires children to think critically and to develop flexible approaches to problem solving scenarios' (Williams et al., 2016, p. 200). Furthermore, Williams et al. (2016) make the following point:

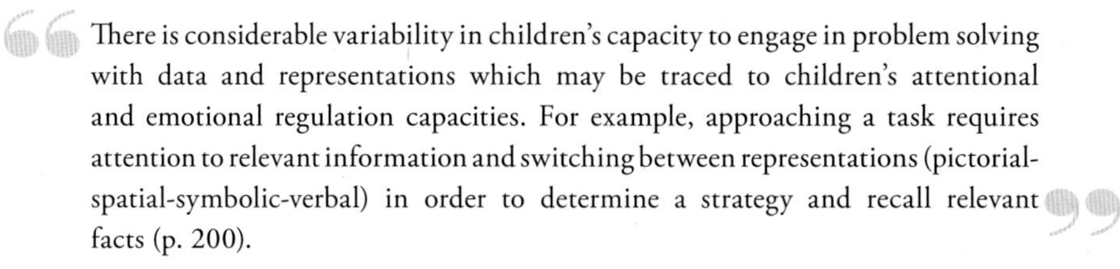

> There is considerable variability in children's capacity to engage in problem solving with data and representations which may be traced to children's attentional and emotional regulation capacities. For example, approaching a task requires attention to relevant information and switching between representations (pictorial-spatial-symbolic-verbal) in order to determine a strategy and recall relevant facts (p. 200).

The research indicates that early childhood education needs to support children's self-regulatory capacities, and that these in turn support children's mathematics achievement. Given the strong relationship between self-regulation and mathematics achievement, Williams et al. (2016) suggest that mathematical learning activities provide the ideal context in which to build children's self-regulatory capacities.

Opinion Piece 3.1

DR CEN (AUDREY) WANG

Learn more
Go online to listen to this Opinion Piece.

I am really glad to see the sharp increase of interest in self-regulation in recent years. It is important that children become competent individuals who are skilful in planning, monitoring, controlling, and reflecting on their cognitions, emotions, behaviours, and motivation. I believe self-regulation is particularly relevant and important in early childhood maths education. Maths tasks usually involve problem solving, which requires children to maintain attention for a period of time as well as shift attention to different aspects of the task when necessary. Children also need to inhibit incorrect responses for a better solution to the problem. Further, maths problems can be challenging at times and children's ability to regulate their affective response is likely to influence their enjoyment of, and persistence in, the task.

Self-regulation skills develop rapidly and can be shaped during the early years. What is more, abundant research has shown that self-regulation skills in the early years are linked to emergent maths skills as well as long-term maths achievement. As such, early childhood maths educators can make a real difference in children's lives by supporting their self-regulatory skills during maths. Making children feel safe and connected, being observant of their behaviour, and identifying triggers of dysregulation—while at the same time acknowledging children's strengths and providing a scaffolding of self-regulation strategies—are good ways to help children develop a sound foundation for maths skills and become self-regulated individuals.

Cen (Audrey) Wang is a Senior Adjunct Research Fellow in the School of Education, Charles Sturt University, Bathurst, Australia. She is particularly interested in educational and developmental psychology. Cen's research focuses on understanding the diverse ways that children learn and promoting children's motivation and social emotional wellbeing. Her recent research explores how contextual factors such as teacher–child relationships, peer relationships, parenting behaviours, and teaching practices contribute to children's self-regulation and social emotional development.

CURRICULUM CONNECTIONS

Early Years Learning Framework

Outcome 1: Children have a strong sense of identity

Children develop their emerging autonomy, inter-dependence, resilience and sense of agency.

This is evident when children:

- Demonstrate an increasing capacity for self-regulation
- Persist when faced with challenges and when first attempts are not successful.

Educators promote this learning when they:

- Provide opportunities for children to engage independently with tasks and play
- Motivate and encourage children to succeed when they are faced with challenges.

Outcome 3: Children have a strong sense of wellbeing

Children become strong in their social and emotional wellbeing.

This is evident when children:

- Make choices, accept challenges, take considered risks, manage change and cope with frustrations and the unexpected
- Show an increasing capacity to understand, self-regulate and manage their emotions in ways that reflect the feelings and needs of others.

Educators promote this learning when they:

- Challenge and support children to engage in and persevere at tasks and play
- Acknowledge and affirm children's effort and growth.

(DET, 2019, p. 25, 34)

Chapter summary

This chapter has canvassed a range of theoretical perspectives that are relevant to, and can inform, early childhood mathematics education. Each of these theories offers different insights into children's mathematical learning and development. No one theory offers a complete view of children's learning and development; rather, knowledge of a range of theories helps us to develop a more comprehensive understanding of how young children develop mathematical knowledge.

The specific topics we covered in this chapter were:

- developmental theories
- constructivism
- social constructivism
- ecological theory
- representation theory
- constructionism
- self-regulation.

FOR FURTHER REFLECTION

1. Which early childhood theories most strongly influence your own thinking and practices?
2. What helps you to relate strongly to these theories?
3. What does your own theoretical preference mean for your teaching of mathematics?
4. How can a range of theories be used to understand children's mathematical learning and development, and to inform your mathematics education practices?

FURTHER READING

Arthur, L., Beecher, B., Death, E., Dockett, S., & Farmer, S. (2015). Thinking about children: Play, learning and development. In *Programming and planning in early childhood settings* (6th ed., pp. 69–106). South Melbourne, VIC: Cengage Learning Australia.

Kearns, K. (2010). Introduction to child development. In *Birth to big school* (pp. 1–72). Frenchs Forest, NSW: Pearson Australia.

Siemon, D., Warren, E., Beswick, K., Faragher, R., Miller, J., Horne, M., Jazby, D., & Breed, M. (2021). *Teaching mathematics: Foundations to middle years* (pp. 32–57). South Melbourne, VIC: Oxford University Press.

Williams, K.E., White, S.L.J., & MacDonald, A. (2016). Early mathematics achievement of boys and girls: Do differences in early self-regulation pathways explain later achievement? *Learning and Individual Differences*, *51*, 199–209.

04 Curricula

Learn more
Go online to see a video from Paige Lee explaining the key messages of this chapter.

CHAPTER OVERVIEW

This chapter explores the roles and function of curricula and highlights the place of mathematics in the two curriculum documents that impact young children in Australia: *Belonging, Being and Becoming: The Early Years Learning Framework for Australia* (EYLF) and the *Australian Curriculum*. The chapter also explores issues of curriculum continuity between early childhood and school settings, and considers the potential for linking the two curricula.

In this chapter, you will learn about the following topics:

- *Early Years Learning Framework for Australia* (EYLF)
- *Australian Curriculum: Mathematics*
- content strands
- proficiency strands
- curriculum continuity
- engaging with curricula

KEY TERMS

- Content strands
- Continuity
- Curriculum/curricula
- Curriculum frameworks
- Outcomes
- Proficiency strands

Introduction

There are varying perspectives on what constitutes a curriculum. A broad definition is that **curriculum** (plural: curricula) refers to everything that happens throughout the day (Arthur et al., 2015). More specifically, a **curriculum framework** provides educators with underpinning principles to guide practice and articulate broad long-term outcomes for children's learning (Arthur et al., 2015). The national early childhood curriculum framework, *Belonging, Being and Becoming: The Early Years Learning Framework for Australia* (EYLF; DET, 2019), takes a view (adapted from *Te Whariki*, the New Zealand national early childhood framework) that an early childhood curriculum refers to 'all the interactions, experiences, activities, routines and events, planned and unplanned, that occur in an environment designed to foster children's learning and development' (p. 9).

Curriculum
refers to everything that happens throughout the day in an educational program.

Curriculum frameworks
guide practice and provide long-term outcomes for children's learning.

In Australia, educational settings for children aged birth to eight years are guided by two national curriculum frameworks: the aforementioned EYLF (DET, 2019) provides a framework for early childhood settings; while primary school programs are informed by the *Australian Curriculum* (ACARA, 2022). As Arthur et al. (2015) explain, 'the EYLF and the Australian Curriculum documents, respectively, specify particular learning outcomes or continuing sequences of learning across learning areas, general capabilities and cross-curricular priorities to which educators must refer when assessing children's learning and documenting their program' (p. 208).

Early Years Learning Framework for Australia (EYLF)

The EYLF is Australia's first national curriculum framework for early childhood education. The EYLF was first developed by the Council of Australian Governments in 2009 'to assist educators to provide young children with opportunities to maximise their potential and develop a foundation for future success in learning' (DET, 2019, p. 5). The EYLF was first published in 2009, and subsequently re-released under Creative Commons license in 2019. The EYLF provides a basis for ensuring that children in early childhood education and care settings have access to quality teaching and learning—in all aspects of their learning, development, and wellbeing.

The EYLF is structured around five **outcomes** for children's learning. The outcomes are framed as the skills, knowledge, and dispositions that educators can actively promote in early childhood settings (DET, 2019, p. 8). The five EYLF outcomes are as follows:

Outcomes
are the skills, knowledge, and dispositions that educators can promote.

1. Children have a strong sense of identity
2. Children are connected with and contribute to their world
3. Children have a strong sense of wellbeing
4. Children are confident and involved learners
5. Children are effective communicators.

What potential for mathematical learning can you identify in the EYLF outcomes?

The EYLF is not intended to be prescriptive; rather, it 'provides broad direction for early childhood educators in early childhood settings to facilitate children's learning' and 'guides educators in their curriculum decision-making' (p. 8). As such, mathematics does not receive 'overt' treatment in the EYLF. However, with careful reading we can see that mathematics is indeed embedded in the EYLF Learning Outcomes in powerful ways.

FIGURE 4.1 Observation record: 'Sandpit play'

Jarvis: Sandpit play

September 5th 2016

Observation

Today Jarvis was playing in the sandpit and filling sand into a large baking tray using a small spoon until it was full and then he tipped it out. When he tipped it out he was laughing. Then he started filling the tray again.

Analysis

Jarvis was holding the spoon with one hand and filling the sand into the tray.

Extension of Learning

Filling and pouring experience for Jarvis as an extension.

EARLY YEARS LEARNING FRAMEWORK

- 1.2: EYLF - Children develop their emerging autonomy, inter-dependence, resilience and sense of agency
- 4.1: EYLF - Children develop dispositions for learning such as curiosity, cooperation, confidence, creativity, commitment, enthusiasm, persistence, imagination and reflexivity
- 4.2: EYLF - Children develop a range of skills and processes such as problem solving, enquiry, experimentation, hypothesising, researching and investigating

Implications for Learning

When he tipped it out he was laughing.

Where to next?

Filling and pouring experience for Jarvis as an extension.

Varinder Kaur
Assistant Educator

(Source: Varinder Kaur/Amy MacDonald)

It is most evident in *Outcome 4: Children are confident and involved learners*. Indeed, Outcome 4 states that:

- Children develop dispositions for learning such as curiosity, cooperation, confidence, creativity, commitment, enthusiasm, persistence, imagination and reflexivity;
- Children develop a range of skills and processes such as problem solving, enquiry, experimentation, hypothesising, researching and investigating;
- Children transfer and adapt what they have learnt from one context to another; and
- Children resource their own learning through connecting with people, place, technologies and natural and processed materials (p. 37).

In early childhood education and care settings, it is expected that educational programs and reports of children's learning are mapped to the EYLF outcomes. As Figures 4.1 and 4.2 demonstrate, the EYLF outcomes can be used to communicate children's mathematical learning as well as the relationship between the mathematical learning and children's more holistic development and wellbeing.

FIGURE 4.2 Observation record: 'Counting on the swing'

Jarvis: Counting on the swing

February 2nd 2017

Observation

Jarvis was on the swing with his friends. With every swing he was counting 1, 2, 3, 4, 5, 6, 7, 8, 9, 10. We repeated this a few times and some of Jarvis' friends joined in.

Extension of Learning

Counting how many steps up to the top of the slide.

EYLF PRACTICES

- 3: PRACTICE - Learning through play

EARLY YEARS LEARNING FRAMEWORK

- 4.3: EYLF - Children transfer and adapt what they have learned from one context to another
- 4: EYLF - Children are confident and involved learners
- 5.1: EYLF - Children interact verbally and non-verbally with others for a range of purposes

Kate Pearce
Assistant Educator

(Source: Kate Pearce/Amy MacDonald)

Australian Curriculum: Mathematics

Primary schools in Australia operate within the context of the *Australian Curriculum* (ACARA, 2022). The role and purpose of the *Australian Curriculum* is articulated by ACARA (2022) as follows:

> The Australian Curriculum sets the expectations for what all Australian students should be taught, regardless of where they live or their background … It means that students have access to the same content, and their achievement can be judged against consistent national standards. Schools and teachers are responsible for the organisation of learning and they will choose contexts for learning and plan learning in ways that best meet their students' needs and interests.

While mathematics receives more subtle treatment in the EYLF, in the *Australian Curriculum* it is relatively easy to appreciate the place of this discipline; indeed, it receives specific attention through the iteration of the curriculum known as the *Australian Curriculum: Mathematics*. The *Australian Curriculum: Mathematics* aims to ensure that students:

- Are confident, creative users and communicators of mathematics, able to investigate, represent and interpret situations in their personal and work lives and as active citizens;
- Develop an increasingly sophisticated understanding of mathematical concepts and fluency with processes, and are able to pose and solve problems and reason in number and algebra, measurement and geometry, and statistics and probability; and
- Recognise connections between the areas of mathematics and other disciplines and appreciate mathematics as an accessible and enjoyable discipline to study (ACARA, 2022).

CONTENT STRANDS

Content strands specify the mathematical topics to be developed.

Mathematics is addressed in the curriculum through three **content strands** that specify the mathematical topics to be developed—these being: Number and Algebra, Measurement and Geometry, and Statistics and Probability. The aims and scope of these three content strands are presented in Table 4.1. The *Australian Curriculum: Mathematics* uses the content of these three strands to articulate an 'achievement standard' for the end of each schooling year. The achievement standards relevant to the early childhood years (Foundation, Year 1, and Year 2) are presented in Table 4.2.

PROFICIENCY STRANDS

Proficiency strands describe how the content is to be explored or developed.

The *Australian Curriculum: Mathematics* also contains four **proficiency strands** that weave across the content strands. These are: Understanding, Fluency, Problem Solving, and Reasoning. As ACARA (2022) explains, 'the proficiencies reinforce the significance of working mathematically within the content and describe how the content is explored or developed. They provide the language to build in the developmental aspects of the learning of mathematics'. In relation to mathematical development in the early years, the document states very specific goals for children's learning each year. By the end of Year 2 (at approximately eight years of age), it is expected that children will have learnt about the following:

> Understanding includes connecting number calculations with counting sequences, partitioning and combining numbers flexibly, identifying and describing the relationship between addition and subtraction and between multiplication and division.
>
> Fluency includes counting numbers in sequences readily, using informal units iteratively to compare measurements, using the language of chance to describe outcomes of familiar chance events and describing and comparing time durations.
>
> Problem Solving includes formulating problems from authentic situations, making models and using number sentences that represent problem situations, and matching transformations with their original shape.

> Reasoning includes using known facts to derive strategies for unfamiliar calculations, comparing and contrasting related models of operations, and creating and interpreting simple representations of data (ACARA, 2022).

Working mathematically through using processes of understanding, fluency, problem solving, and reasoning is explored further in Chapter 9.

TABLE 4.1 Content strands of the *Australian Curriculum: Mathematics*

NUMBER AND ALGEBRA	Number and Algebra are developed together, as each enriches the study of the other. Students apply number sense and strategies for counting and representing numbers. They explore the magnitude and properties of numbers. They apply a range of strategies for computation and understand the connections between operations. They recognise patterns and understand the concepts of variable and function. They build on their understanding of the number system to describe relationships and formulate generalisations. They recognise equivalence and solve equations and inequalities. They apply their number and algebra skills to conduct investigations, solve problems and communicate their reasoning.
MEASUREMENT AND GEOMETRY	Measurement and Geometry are presented together to emphasise their relationship to each other, enhancing their practical relevance. Students develop an increasingly sophisticated understanding of size, shape, relative position and movement of two-dimensional figures in the plane and three-dimensional objects in space. They investigate properties and apply their understanding of them to define, compare and construct figures and objects. They learn to develop geometric arguments. They make meaningful measurements of quantities, choosing appropriate metric units of measurement. They build an understanding of the connections between units and calculate derived measures such as area, speed and density.
STATISTICS AND PROBABILITY	Statistics and Probability initially develop in parallel and the curriculum then progressively builds the links between them. Students recognise and analyse data and draw inferences. They represent, summarise and interpret data and undertake purposeful investigations involving the collection and interpretation of data. They assess likelihood and assign probabilities using experimental and theoretical approaches. They develop an increasingly sophisticated ability to critically evaluate chance and data concepts and make reasoned judgements and decisions, as well as building skills to critically evaluate statistical information and develop intuitions about data.

(Source: ACARA, 2022)

TABLE 4.2 *Australian Curriculum: Mathematics* achievement standards for Foundation, Year 1, and Year 2

FOUNDATION	YEAR 1	YEAR 2
By the end of the Foundation year, students make connections between number names, numerals and quantities up to 10. They compare objects using mass, length and capacity. Students connect events and days of the week. They explain the order and duration of events. They use appropriate language to describe location. Students count to and from 20 and order small collections. They group objects based on common characteristics and sort shapes and objects. Students answer simple questions to collect information and make simple inferences.	By the end of Year 1, students describe number sequences resulting from skip counting by 2s, 5s and 10s. They identify representations of one half. They recognise Australian coins according to their value. Students explain time durations. They describe two-dimensional shapes and three-dimensional objects. Students describe data displays. Students count to and from 100 and locate numbers on a number line. They carry out simple additions and subtractions using counting strategies. They partition numbers using place value. They continue simple patterns involving numbers and objects. Students order objects based on lengths and capacities using informal units. They tell time to the half-hour. They use the language of direction to move from place to place. Students classify outcomes of simple familiar events. They collect data by asking questions, draw simple data displays and make simple inferences.	By the end of Year 2, students recognise increasing and decreasing number sequences involving 2s, 3s and 5s. They represent multiplication and division by grouping into sets. They associate collections of Australian coins with their value. Students identify the missing element in a number sequence. Students recognise the features of three-dimensional objects. They interpret simple maps of familiar locations. They explain the effects of one-step transformations. Students make sense of collected information. Students count to and from 1000. They perform simple addition and subtraction calculations using a range of strategies. They divide collections and shapes into halves, quarters and eighths. Students order shapes and objects using informal units. They tell time to the quarter-hour and use a calendar to identify the date and the months included in seasons. They draw two-dimensional shapes. They describe outcomes for everyday events. Students collect, organise and represent data to make simple inferences.

(Source: ACARA, 2022)

Curriculum continuity

The EYLF and the *Australian Curriculum: Mathematics* were developed at different times, by different groups of people; and as such, the different structures, content, and philosophies of the two documents can make it difficult to see links between them (MacDonald, Goff, Dockett, & Perry, 2016). As such, a challenge for educators is to create appropriate continuity for children as they transition from one curriculum to the other (Perry, Dockett, & Harley, 2012). **Continuity** can be considered as a sense of connection from one context to the next, in which knowledge, skills, and approaches are recognised and transferred across settings.

Continuity means a sense of connection from one context to the next.

Opinion Piece 4.1

Learn more
Go online to listen to this Opinion Piece.

DR JESSAMY DAVIES

As part of my doctoral research, I explored the ways in which educators' pedagogical practices could enhance the transition to school experiences for children, their families, and other stakeholders through the use of educator networks. One of the ways that transition to school can be enhanced is through continuity of pedagogy from prior-to-school to school settings. Of particular relevance in the field of early childhood are the curriculum documents: the *Early Years Learning Framework* and the *Australian Curriculum*. The results of my research indicate that while the curriculum documents do not, in themselves, align (and are not intended to), it is the pedagogy of the educators that can provide continuity across different settings. Therefore, it is the educators who are fundamental to supporting children's continuity of learning.

Jessamy Davies is a Lecturer in Education at Charles Sturt University, Albury–Wodonga, Australia. Her doctoral work was undertaken as part of an Australian Research Council Discovery project that explored policy–practice trajectories at the time of transition to school. The project examined the policy intentions and impact of the *Early Years Learning Framework* and the *Australian Curriculum* on transition to school at national, state, and local levels.

Perry et al. (2012) suggest that one way of creating continuity between the curricula in early childhood and school settings is to consider the powerful mathematical ideas that are experienced by children in both settings. They have developed a *Numeracy Matrix* (Perry et al., 2012) which provides a mechanism of linking the EYLF outcomes with the *Australian Curriculum: Mathematics* strands and proficiencies. An example from this *Numeracy Matrix* is presented in Table 4.3. The *Numeracy Matrix* is based on the use of 'pedagogical inquiry questions', which, as Perry et al. explain, 'means that early childhood educators can be asking themselves the same questions about their own pedagogical practice but answering them through links to whichever of the two curriculum documents pertains to their setting' (p. 161). While the learning opportunities and pedagogical approaches may differ between early childhood and school settings, recognising children's knowledge of, and engagement with, powerful mathematical ideas may help to facilitate a continuum of learning and continuity across settings. Moreover, attention to the mathematical ideas being explored by children can help to create common language and constructs in order for early childhood and school educators to communicate about children's mathematical learning (Perry et al., 2012).

TABLE 4.3 Example from the *Numeracy Matrix*

EYLF OUTCOMES/ AUSTRALIAN CURRICULUM: MATHEMATICS STRANDS AND PROFICIENCIES	Outcome 1: Children have a strong sense of identity
NUMBER AND ALGEBRA	What opportunities do we provide for children to seek new challenges and persist in their problem solving? In what ways do we assist children to represent varied physical activities and games through patterns and symbols? How do we encourage children to explore different perspectives in mathematical problem solving?
MEASUREMENT AND GEOMETRY	How do we encourage children to work collaboratively with peers during measurement activities? In what ways are children able to demonstrate flexibility and make choices when playing with collections of everyday shapes and objects?
STATISTICS AND PROBABILITY	How do we encourage children to develop a notion of fairness in their lives?
UNDERSTANDING	How do we encourage children to play and interact purposefully with the mathematics they experience in their lives? What do we do to assist children link real world representations of mathematics with their own mathematical words and symbols?
FLUENCY	How do we encourage children to use different communication strategies to organise and clarify their mathematical thinking? What opportunities do children have to show their patterning abilities?
PROBLEM SOLVING	How do we encourage children to use the process of play, reflection and investigation to solve mathematical problems? How do we assist children to gain confidence in their ability to explore, hypothesise and make appropriate choices in their mathematics?
REASONING	How do we encourage children to demonstrate flexibility and to manage different mathematical ideas as they are presented to them by peers?

(Source: Perry, Dockett & Harley, 2012, p. 162)

Engaging with curricula

As indicated in Chapter 1, 'Curriculum connections' features throughout this text provide links between chapter content and the EYLF and the *Australian Curriculum: Mathematics*. These features are intended as a guide for how the mathematical content and pedagogical approaches explored in this text align to the curricula with which early years educators work. However, approaches to curriculum are rarely fixed and straightforward; rather, curricula operate at many points along continuums that can be used flexibly to guide educational programs (Arthur et al., 2015). Moreover, the enactment of curriculum includes leadership, pedagogy and knowledge, interactions between children and educators, tools and resources, and the actual mathematics presented (Way et al., 2016).

Your capacity to engage effectively with curricula will be enhanced by having a comprehensive understanding of mathematics as a discipline and of children's developing mathematical identities. I encourage you to interpret, and value-add to, early years curriculum frameworks in ways that make sense for you and the children with whom you work.

What do you see as the role of curricula in your early childhood mathematics education practice? How might you use curriculum documents to inform your planning and pedagogical approaches?

Chapter summary

This chapter has presented the curriculum foundations for early childhood mathematics education. Both the *Early Years Learning Framework for Australia* and the *Australian Curriculum: Mathematics* provide guidance for early childhood mathematics education; albeit in very different ways. Knowledge of both curricula can help educators to support continuity of children's mathematics education as they transition from prior-to-school to school contexts.

The specific topics we covered in this chapter were:

- *Early Years Learning Framework for Australia* (EYLF)
- *Australian Curriculum: Mathematics*
- content strands
- proficiency strands
- curriculum continuity
- engaging with curricula.

FOR FURTHER REFLECTION

1 What points stood out to you about the curriculum foundations for early childhood mathematics education?

2 Were there common elements between the EYLF and the *Australian Curriculum: Mathematics?* In what ways might these curricula be connected?

3 How might continuity between these curricula be created within an early childhood mathematics education program?

FURTHER READING

Arthur, L., Beecher, B., Death, E., Dockett, S., & Farmer, S. (2015). Curriculum approaches and pedagogies. In *Programming and planning in early childhood settings* (6th ed., pp. 207–56). South Melbourne, VIC: Cengage Learning Australia.

Australian Curriculum, Assessment and Reporting Authority. (2022). *Australian Curriculum: Mathematics (Version 8.4)*. Available online: https://www.australiancurriculum.edu.au/f-10-curriculum/mathematics/

Australian Government Department of Education and Training. (2019). *Belonging, being and becoming: The Early Years Learning Framework for Australia*. Available online: https://www.education.gov.au/child-care-package/resources/belonging-being-becoming-early-years-learning-framework-australia

Perry, B., Dockett, S., & Harley, E. (2012). The Early Years Learning Framework for Australia and the Australian Curriculum: Mathematics—Linking educators' practice through pedagogical inquiry questions. In B. Atweh, M. Goos, R. Jorgensen, & D. Siemon (eds), *Engaging the Australian National Curriculum: Mathematics—Perspectives from the field* (pp. 155–174). Mathematics Education Research Group of Australasia. Available online: https://merga.net.au/Public/Public/Publications/Engaging_the_Australian_curriculum_mathematics_book.aspx

Part 2

APPROACHES TO TEACHING MATHEMATICS IN EARLY CHILDHOOD EDUCATION

OVERVIEW OF PART 2

Part 2 of this book consists of seven chapters that examine different approaches to teaching mathematics in early childhood education. Individually, the chapters address the following topics:

» educator roles in mathematics education
» involving families in mathematics education
» strengths approaches to mathematics education
» informal opportunities to learn about mathematics
» working mathematically through powerful processes
» mathematics and transitions to school
» the relationship between assessment and planning.

Collectively, these chapters focus on the different ways of exploring mathematics with young children in a range of early childhood education contexts, and different pedagogical considerations and strategies. Part 2 provides a toolkit for your role as an early childhood mathematics educator and presents a variety of pedagogies and practices for you to draw upon. The information in these chapters will assist you to become a thoughtful and responsive mathematics educator who takes into account the needs, interests, perspectives, and values of the different stakeholders in early childhood mathematics education.

05 Educator Roles

Learn more
Go online to see a video from Paige Lee explaining the key messages of this chapter.

CHAPTER OVERVIEW

This chapter will explore the different pedagogical actions that early childhood educators can use to engage children with explorations of mathematics. The chapter also examines the role of educators as 'enculturators' of mathematics who introduce young children to social and cultural perspectives of mathematics. This chapter provides explanation of these different pedagogical practices, and examples of when and how they might be utilised.

In this chapter, you will learn about the following topics:

- educators as mathematical enculturators
- playful pedagogical approaches
- child-instigated and educator-instigated experiences
- indirect and direct teaching
- intentional teaching
- responsive strategies
- listening and observing
- responding
- questioning
- collaborating
- modelling
- scaffolding
- communication roles
- the role of reflective practice
- shared contexts and shared responsibilities.

KEY TERMS

- Child-instigated experiences
- Direct teaching
- Educator-instigated experiences
- Enculturation
- Evaluator role
- Indirect teaching
- Instructor role
- Intentional teaching
- Mediator role
- Modelling
- Playful pedagogies
- Reflective practice
- Responding
- Responsive strategies
- Restrictive strategies
- Shared contexts

Introduction

Bishop (1991) argued that mathematics is a 'way of knowing'. He reinforced the notion that mathematics is a cultural product that embodies cultural values, and that environmental and societal activities stimulate mathematical concepts. However, Bishop also emphasised that 'people are responsible for the process' (p. 160). That is—educators have a responsibility for children's mathematical **enculturation**:

Enculturation is a process by which people learn the values and behaviour appropriate to a culture.

> The process is at heart an interpersonal one—we must emphasise the role of *people* and particularly those who have special responsibilities for this process … A mathematics teacher, in her role as mathematical enculturator, interacts with and engages with mathematical culture in a different way from others and therefore she needs to know mathematical culture in a different way … She needs to know it both globally and locally … She needs to know the relationships … to understand its actual and potential contribution to society (pp. 160–2).

Educators can employ specific pedagogical actions that welcome children into mathematical worlds and help children to draw mathematical meaning from activities in their everyday personal and social environments. This chapter articulates the different pedagogical roles and actions of early childhood mathematics educators that help to enculturate children into meaningful mathematical worlds. It draws upon the seminal work of Agnes Macmillan (2009), which positioned early childhood mathematics education as a *shared context* in which educators could use thoughtful pedagogical actions to engage children with social and cultural perspectives of mathematics.

Opinion Piece 5.1

DR SHEENA ELWICK

Learn more
Go online to listen to this Opinion Piece.

Agnes Macmillan and I met while I was studying my Bachelor of Education (Early Childhood and Primary) (Honours) and it was then that Agnes both inspired my love of teaching mathematics to young children, and contributed to my understandings of how to do so responsively. That was no mean feat. Prior to meeting Agnes, I was convinced that mathematics was a subject that had little relevance to everyday life and that children did not learn mathematics until they started primary school. It was only then that children became competent (or not) at 'doing' mathematics. The idea that children enter primary school as highly competent individuals with a natural learning potential for mathematics, and a wealth of mathematical knowledge honed by everyday life, never entered my mind. How strange that oversight seems now! Agnes not only helped me to recognise the many ways in which young children explore and come to know mathematical meanings as they play, talk, and carry out their everyday lives, she also gave me the language to discuss those mathematical meanings with others. I have no doubt that it was through Agnes's explicit teaching and provision of many opportunities for applying Bishop's (1988) 'Six Universal Mathematical Activities' to practice that I will never see children's play as 'just play' ever again.

I have no doubt, also, that I am not alone in thinking this way. Little did Agnes and I know, but our meeting would be the start of a friendship and collegial relationship that would span such a length of time that I am now able to share Agnes's work with the many pre-service early childhood teachers with whom I have the pleasure of working. Agnes's book *Numeracy in Early Childhood* has formed the basis of much of that work, and comments such as 'oh, I thought they were just playing' are commonly heard during my workshops. Perhaps the most memorable comment, though, was one made recently by a student who regularly 'babysat' children for one of her relatives. Suddenly, part-way through a discussion of Agnes's description of Bishop's mathematical processes, this student exclaimed: 'But I thought I was just babysitting. I never knew that when I cooked cupcakes with the children I was providing them with so many opportunities to explore mathematics!' Her excitement was infectious and prompted a lengthy discussion during which she, and other students, considered their own roles and potential expertise in cultivating young children's learning of mathematics during everyday moments such as cooking. To be able to share Agnes's work in this way, and to inspire future early childhood teachers in the way that Agnes inspired me is, without doubt, one of the most rewarding aspects of my career.

Sheena Elwick is a Senior Lecturer in Early Childhood Education at Charles Sturt University, Wagga Wagga, Australia. Sheena has a broad range of experience in primary school and early childhood education and care settings, both as a practitioner and as a teacher educator. As a practitioner, she has worked in long day care, mobile childcare, and also primary school settings in both New South Wales and Victoria. She has also held several leadership positions in each of these settings, including Mathematics Coordinator (F–6) and Common Curriculum Numeracy Team Leader (F–12).

Educators as mathematical enculturators

In enacting the roles of an early childhood mathematics educator, we serve to become a mathematics *enculturator* for the young children with whom we work. Bishop (1991, pp. 164–8) outlined several qualities of mathematical enculturators:

- *Ability to personify mathematical culture*—a mathematics educator is not just someone who is proficient in mathematics; a mathematics educator 'models' the techniques, feelings, ideologies, and behaviours of the culture of mathematics.
- *Commitment to the mathematical enculturation process*—the responsibilities of a mathematics educator can be morally demanding; an educator must always have the mathematics learner as their primary focus, with respect for the learner's behaviours and feelings, which are being shaped.
- *Ability to communicate mathematical ideas and values*—mathematical enculturation is a special kind of communication process, and mathematics educators need to enjoy and be successful at this communication.
- *Accountability to the mathematical culture*—a mathematics educator has responsibility *to* mathematics and for the preservation and growth of mathematical culture.

What was your experience of being enculturated to mathematics? How were you supported to access the concepts, processes, and values of mathematics? Did your experiences generate positive or negative feelings towards mathematics?

In her 2009 text, Macmillan extended the work of Bishop by bringing to bear a socio-cultural view of mathematics education for early childhood contexts. Macmillan emphasised the importance of children's *numeracy*, which she conceptualised as 'a social and cultural perspective for discovering and thinking about mathematical knowledge and applying it to fulfil the purposes of our everyday lives' (p. 1). Furthermore, Macmillan argued that numeracy processes are present and functioning in children's play contexts, and that educators can employ pedagogical actions that support children's numeracy enculturation through play situations.

Playful pedagogical approaches

It is well established that play and learning are inextricably woven together. Play helps children become abstract thinkers; it makes the abstract have form and makes the intangibleness of ideas tangible and manageable (Bruce, 2011). Play is intrinsically motivating, and provides the foundation for social, emotional, cognitive, and physical development in the early childhood years (Ebbeck & Waniganayake, 2010). When children are playing, they are exploring and investigating objects and meanings, and engaging in many different thinking processes (Macmillan, 2009). As Bruce (2011) explains, through their play children, for example,

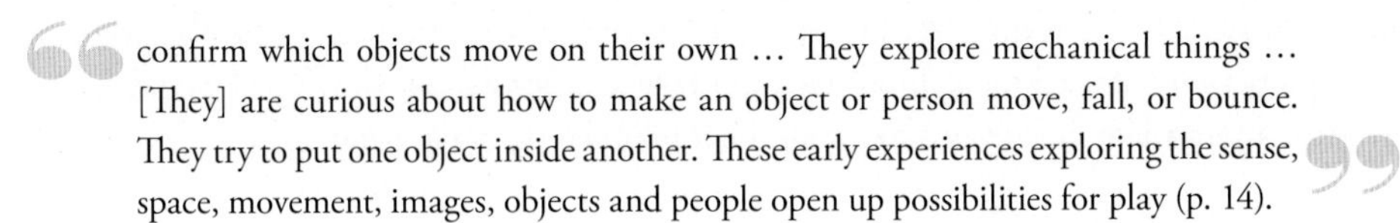

> confirm which objects move on their own … They explore mechanical things … [They] are curious about how to make an object or person move, fall, or bounce. They try to put one object inside another. These early experiences exploring the sense, space, movement, images, objects and people open up possibilities for play (p. 14).

Play provides meaningful personal and social contexts in which to explore mathematical concepts and applications. In a previous text (MacDonald, 2015a), I advocated for **playful pedagogies**—that is, pedagogical approaches that are playful for all involved, adult and child alike. Playful pedagogies both require, and facilitate, the curiosity, creativity, flexibility, and adaptability of all players. They are particularly important in mathematics education as they open up the possibility of the unstructured and the unknown—notions that underpin problem posing and problem solving in mathematics.

Playful pedagogies are approaches that are playful for all involved (adults and children).

As Hunting, Mousley, and Perry (2012) explain, play is a frame of mind and a way of engaging—and as such, authentic playfulness also requires *mindfulness* and *responsiveness* to the circumstances that present themselves. We need to have a robust understanding of the mathematical concepts and processes that might be embedded in the play situation, and have the capacity, where appropriate, to take the play to new and different places so as to challenge children's existing understandings and generate new knowledge.

Knowing when to 'step in' also means knowing when to 'pull back' and allow the play to take its own direction. Adults need to be sensitive in play situations (Bruce, 2011). There can be a subtle distinction between being a 'play-designer' and a 'play-participant', and both adults and children can perform both roles. Moreover, a playful pedagogical approach makes room for both.

Child-instigated and educator-instigated experiences

There are two distinct approaches to planning for play and investigation in mathematics education: child-instigated experiences and educator-instigated experiences (Harlan & Rivkin, 2012). This could also be conceptualised as 'spontaneous' versus 'planned' learning experiences (Arthur et al., 2015). As Harlan and Rivkin (2012) explain, **child-instigated experiences** are incidental and can occur at any time or place—whenever a child's curiosity is sparked. The educator capitalises on the child's curiosity by asking questions that lead to further discovery, by relating the interest to something the child already knows, by extending the interest to other activities, or by offering to help the child locate other resources to expand their interest and understanding (Harlan & Rivkin, 2012).

Child-instigated experiences are incidental learning experiences led by the child's curiosity.

Educator-instigated experiences are learning experiences that are planned in advance and are directed by the educator.

On the other hand, **educator-instigated experiences** (or educator-guided instruction, as it is often known) do not rely exclusively on child-initiated experiences to promote understandings of mathematics (Copley et al., 2007). Rather, learning opportunities—which are planned in advance—are designed and provided by the educator, and introduce concepts and terminology that children may not be able to discover on their own (Copley et al., 2007).

It is important to take a balanced approach to mathematics education, utilising both child-instigated and educator-instigated experiences.

Planning for specific experiences based on children's previously identified interests, emerging interests, and ongoing investigations or projects requires educators to be both highly organised and flexible (Arthur et al., 2015). Educators also need to be able to draw upon a repertoire of teaching approaches and make decisions about which pedagogical roles are most appropriate in a given situation—be it planned or spontaneous.

Indirect and direct teaching

When considering the educator's role in mathematical learning, it is important to think about whether the approach is *indirect* or *direct*. Children's play and investigation can be guided by **indirect teaching** through the use of thoughtful questioning and listening and by the sensitive leading of discussions (Harlan & Rivkin, 2012). When adults are curious investigators *with* children, they can help to cultivate children's natural curiosity and interests (Macmillan, 2009).

Indirect teaching involves guiding play and investigation through thoughtful questioning and listening.

Direct teaching involves offering conceptual cues and suggesting more effective strategies.

Children's play and investigation might also be guided by **direct teaching**—that is, by offering conceptual cues and suggesting more effective strategies (Harlan & Rivkin, 2012). Macmillan (2009, p. 23) noted that direct teaching in investigatory contexts requires educators to be aware of opportunities for:

- encouraging problem-solving
- observing changing and unchanging phenomena in specific detail, using all the senses
- interacting using explicit language and providing opportunities for documenting
- reporting and talking about discoveries and solutions to problems.

INTENTIONAL TEACHING

It is important that educators are thoughtful and intentional in the pedagogies that they choose (Arthur et al., 2015). The term **intentional teaching** is used to highlight educators as active and purposeful in selecting pedagogies that are most appropriate to the learning situation (Arthur et al., 2015). The *Early Years Learning Framework for Australia* (EYLF; DET, 2019) advocates for intentional teaching practices to foster children's learning in deliberate, purposeful, and thoughtful ways:

Intentional teaching refers to being active and purposeful in selecting pedagogies.

> Educators who engage in intentional teaching recognise that learning occurs in social contexts and that interactions and conversations are vitally important for learning … Educators move flexibly in and out of different roles and draw on different strategies as the context changes (p. 18).

Within an intentional teaching framework, both indirect and direct pedagogical approaches are necessary for guiding children's learning of mathematical concepts and processes and responding to spontaneous learning opportunities that arise.

Responsive strategies

Macmillan (2009) suggests that teaching practices can be either *restrictive* or *responsive.* **Restrictive strategies** restrict children's access to mathematical knowledge and deny equitable relations of control and participation. According to Macmillan (2009, p. 152), restrictive strategies are likely to involve language or approaches that:

Restrictive strategies restrict access to mathematical knowledge and create inequitable control and participation.

- deny children access to mathematical content because its meanings or purposes are imposed, unclear, confusing, inaccessible, non-explanatory, or dominated by instruction
- create too great a challenge for the child because the concepts or procedures are too difficult
- place too much emphasis on evaluation
- deny access to assistance, support, and feedback, and/or
- restrict opportunities to negotiate the learning opportunities.

On the other hand, **responsive strategies** provide access to mathematical knowledge and generate respect, recognition, and cooperation. Responsive strategies consider the ways in which children can access mathematical concepts, understandings, thinking processes, and procedures, and focus on fostering positive feelings towards mathematics. Responsive educators check that the purposes and processes for a particular mathematical experience are clear and accessible, and they ensure that there are opportunities for children to negotiate their way towards new or more developed mathematical understandings (Macmillan, 2009).

Responsive strategies provide access to mathematical knowledge and generate respect, recognition, and cooperation.

Pause and reflect

Think about the strategies you use in your interactions with young children. Would you describe these as responsive or restrictive? In what ways might you adapt your communication style to become more responsive to children?

Responsive strategies are valued within the EYLF (DET, 2019), which positions *responsiveness to children* as a key pedagogical practice to promote children's learning:

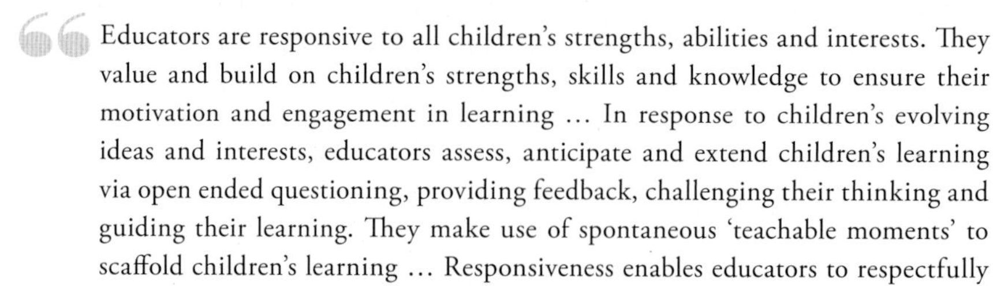

> Educators are responsive to all children's strengths, abilities and interests. They value and build on children's strengths, skills and knowledge to ensure their motivation and engagement in learning … In response to children's evolving ideas and interests, educators assess, anticipate and extend children's learning via open ended questioning, providing feedback, challenging their thinking and guiding their learning. They make use of spontaneous 'teachable moments' to scaffold children's learning … Responsiveness enables educators to respectfully enter children's play and ongoing projects, stimulate their thinking and enrich their learning (pp. 14–15).

Key teaching strategies that generate responsiveness include *listening and observing*, *responding*, *questioning*, *collaborating*, *modelling*, and *scaffolding*. These strategies are discussed in the following sections.

LISTENING AND OBSERVING

Listening to, and observing, children's mathematical explorations and ideas requires educators to attend to, analyse, interpret, and assess what they see and hear (Macmillan, 2009). Taking the time to listen carefully to children's conversations and explanations, and also watch closely as children carry out thoughtful and purposeful actions, can offer great insight into children's mathematical thinking. The information we see and hear can provide us with entry points into children's mathematical investigations, provide evidence upon which to base our pedagogical decision making, and enable us to respond to children's mathematical meaning-making in appropriate ways.

RESPONDING

Responding refers to the use of statements or questions that show understanding or invite responses.

The pedagogical approach of **responding** refers to the use of statements or questions that show you have understood the child's contribution or that invite a response from the child (Macmillan, 2009). Responsive statements provide encouragement, interest, inquiry, or confirmation—for example, 'I see that you're filling the bottles right to the top' (Macmillan, 2009, p. 165). Responsive educators offer encouragement as well as challenge, and engage in discussions (Arthur et al., 2015).

QUESTIONING

Questioning as a pedagogical approach can help to identify and extend children's mathematical thinking and knowledge. Metacognitive questions such as 'What do you mean?' and 'Why do you say that?' provide a framework for children to think about their responses (Siemon et al., 2021, p.106). Careful questioning can encourage children to articulate *how* and *why* they have engaged with a mathematical idea or situation in a particular way. Questioning to encourage mathematical thinking is explored in more detail in Chapter 9.

COLLABORATING

Collaboration requires educators and children to take on more or less equal roles in the learning situation. Collaborative learning experiences support children to access ideas and undertake tasks that may be too challenging if attempted without support. Moreover, collaboration allows educators to learn about children's mathematical thought processes as they engage in a shared activity. Macmillan (2009, p. 156) suggests the following techniques for encouraging collaboration:

- Create mutuality by using first-person plural pronouns (for example, 'we', 'us', 'our').
- Play and engage with the children in parallel play or shared activity.
- Pretend that you don't know, so that children feel willing to contribute and take charge.
- Introduce probability (for example, 'We could do it one way, or we could do it another—which would be best?') to encourage collaborative investigations.

CURRICULUM CONNECTIONS

Early Years Learning Framework

Outcome 1: Children have a strong sense of identity

Children feel safe, secure, and supported.

This is evident when children:

- Establish and maintain respectful, trusting relationships with other children and educators
- Openly express their feelings and ideas in their interactions with others
- Respond to ideas and suggestions from others
- Initiate interactions and conversations with trusted educators.

Educators promote this learning when they:

- Acknowledge and respond sensitively to children's cues and signals
- Spend time interacting and conversing with each child.

(DET, 2019, p. 24)

MODELLING

Modelling is a responsive strategy that provides opportunities for learners to model, test, and check all kinds of mathematical knowledge (Macmillan, 2009). Modelling may take verbal or non-verbal forms, and the modelling may be performed by the educator or the children. When using modelling as a teaching strategy, educators may demonstrate how specific language or terminology is used, utilise concrete materials to demonstrate abstract ideas, or model curiosity and reflection techniques (Macmillan, 2009). Similarly, educators should encourage children to construct their own models as a way of representing and testing out their mathematical thinking.

Modelling provides opportunities to model or exemplify, test, and check mathematical knowledge.

CURRICULUM CONNECTIONS

Early Years Learning Framework

Outcome 4: Children are confident and involved learners.

Children develop a range of skills and processes such as problem solving, inquiry, experimentation, hypothesising, researching and investigating.

This is evident when children:

- Contribute constructively to mathematical discussions and arguments.

Educators promote this learning when they:

- Intentionally scaffold children's understandings.

Children resource their own learning through connecting with people, place, technologies and natural and processed materials.

This is evident when children:

- Engage in learning relationships
- Experience the benefits and pleasures of shared learning exploration.

Educators promote this learning when they:

- Provide opportunities and support for children to engage in meaningful learning relationships
- Think carefully about how children are grouped for play, considering possibilities for peer scaffolding.

(DET, 2019, p. 38, 40)

SCAFFOLDING

As described in Chapter 3, scaffolding is a process of providing guidance and support to children moving from one level of competence to another, matching the amount of support to the skill level the learners display (Arthur et al., 2015). Within mathematics education, scaffolding may look different at different times. For example, an educator may scaffold a child to construct a meaningful visual representation (such as a drawing or construction) of a concept they have been exploring. The educator may assist the child to transfer between representational forms, such as by drawing something they have constructed. They may help to scaffold the child's mathematical language by modelling appropriate terminology to describe a mathematical concept the child has been investigating. These are but a few examples, but they demonstrate the diverse ways in which an educator might respond to a child's mathematical thought and action—scaffolding the child so as to extend and enrich their learning experience through exposure to different ways of thinking and acting.

CURRICULUM CONNECTIONS

Early Years Learning Framework

Outcome 5: Children are effective communicators

Children interact verbally and non-verbally with others for a range of purposes.

This is evident when children:

- Interact with others to explore ideas and concepts, clarify and challenge thinking, negotiate and share new understandings
- Demonstrate an increasing understanding of measurement and number using vocabulary to describe size, length, volume, capacity and names of numbers
- Use language to communicate thinking about quantities to describe attributes of objects and collections, and to explain mathematical ideas.

Educators promote this learning when they:

- Model language and encourage children to express themselves through language in a range of contexts and for a range of purposes
- Engage in sustained communication with children about ideas and experiences, and extend their vocabulary
- Include real-life resources to promote children's use of mathematical language.

(DET, 2019, p. 43)

Communication roles

One of the most influential aspects of Macmillan's (2009) work has been her articulation of the different communication roles that early childhood educators can employ to foster children's mathematical learning. As Macmillan explains:

> When we communicate we choose structures that are appropriate in a given context for fulfilling particular needs, wants, goals or responsibilities ... The context determines, to a large extent, the kinds of content and language structures that are used. Personal and social purposes and responsibilities combine with context to determine the appropriate kinds of texts and kinds of *communication roles* to adopt ... That is to say, educators adopt communication roles and strategies to suit the needs and purposes both of the community they represent and of the learner (pp. 50–1).

Instructor role focuses on providing and managing knowledge-acquisition processes.

Mediator role focuses on shared meaning-making and facilitation of learning.

Evaluator role focuses on analysing, interpreting, assessing, and providing feedback.

Macmillan describes three communication roles that are relevant in early childhood mathematics education contexts: **instructor**, **mediator**, and **evaluator**. An overview of these three roles is presented in Table 5.1.

Instructor and *evaluator* roles may function as one- or two-way communications and have the potential to generate responsive or restrictive power and knowledge relations (Macmillan, 2009). On the other hand, the *mediator* role is a two-way process that connects

TABLE 5.1 Overview of communication roles

ROLES	RESPONSIBILITIES	PURPOSES
Instructor: a one- or two-way role	To instruct by explaining the who, what, when, where, why, and how of things said and done, and how they can be understood	To provide and manage the processes involved in knowledge acquisition To set up and monitor the physical, emotional, social, and cultural dimensions of the environment
Mediator: a two-way role	To share the meaning-making process as facilitator of learning	To collaborate through shared: » language » control » participation » responsibility » resources
Evaluator: a one- or two-way role	To analyse, interpret, assess, and judge: » conceptual or procedural knowledge » perceptual, physical, psychological, and emotional factors » social and cultural factors	To provide feedback that is: » responsive » relevant » purposeful » constructive » insightful

(Adapted from Macmillan, 2009, pp. 50–1)

one set of knowledge with another—the mediator role makes connections between what is known and what is new (Macmillan, 2009). Macmillan provides a more detailed explanation of the mediator role:

> In the mediator role, educators *listen* to the learner's thoughts, *observe* their actions on objects, *interpret* them, and use this interpretation to *respond* with a comment or question. Mediating strategies model, guide, instruct, facilitate and evaluate the thinking processes occurring in a particular context. Learners and educators have access to each other's thought processes and this access informs them of each other's progress (Clarke, Clarke, & Cheeseman, 2006) ... The mutually constructed nature of the process creates shared contexts for teaching and learning (pp. 51–2).

Responsive pedagogical strategies are most easily generated when educators take on a mediator role, as there is a shared control over the production of knowledge and artefacts (Macmillan, 2009). The role of mediator is particularly helpful for young children as it generates optimum opportunities for their needs and interests to be supported by peers and knowledgeable others (Macmillan, 2009), which is consistent with Vygotsky's concept of scaffolding within the Zone of Proximal Development. Enacting a mediator role enables the co-construction of mathematical knowledge in ways that are meaningful to all participants in the learning situation.

Think about the mathematics educators you have encountered in your own educational experiences. What roles were taken by these educators—instructor, mediator, or evaluator? How did these different role types influence your feelings about mathematics?

The role of reflective practice

Reflective practice is one of five principles underpinning the EYLF (DET, 2019). It is described in the EYLF as:

> a form of ongoing learning that involves engaging with questions of philosophy, ethics and practice. Its intention is to gather information and gain insights that support, inform and enrich decision-making about children's learning. As professionals, early childhood educators examine what happens in their settings and reflect on what they might change. Critical reflection involves closely examining all aspects of events and experiences from different perspectives (p. 14).

Reflective practice involves examining what happens in educational settings and reflecting on what might be changed.

Reflection can occur at different times in different ways. *Reflection in action* is about constantly reflecting and making decisions about how to respond to children and how to implement spontaneous experiences (Arthur et al., 2015). *Reflection on action* can occur before a learning experience is implemented as part of the planning process, or after an experience or event (Arthur et al., 2015). Reflective practice may involve the support of a mentor or peer, or collaboration with colleagues. For reflection to be productive, educators need to think about how things could be improved, how challenges could be solved, and what specific strategies could be explored in the future (Macmillan, 2009).

Shared contexts and shared responsibilities

In her 2009 book, Macmillan positioned early childhood mathematics education as a *shared context* for teaching and learning. **Shared contexts** can be considered as interactive spaces that require 'equity, balance and flexibility in the process of participation, making meanings and producing knowledge' (Macmillan, 2009, p. 131). Within this shared context, mathematical meanings and understandings are negotiated among educators and children, and educators employ thoughtful decision making about the roles they enact and pedagogical approaches they use.

Shared contexts consist of equity, balance, and flexibility in participation, meaning-making, and knowledge production.

We, as mathematics educators, have a shared responsibility for the ways in which children experience mathematics education and acquire mathematical understandings. Bishop (1991) reinforces this notion of collective responsibility for mathematical enculturation:

> We who are a part of the mathematics education community are all in some way responsible for the mathematical enculturation process. We can take pleasure from its successes and we must carry the blame when it is not so successful ... The mathematics education community must help to create the favourable climate and context within which that formal enculturation can happen ... Moreover we should strive within our own cultural community to demonstrate as much as possible the values of mathematical culture (pp. 177–8).

Chapter summary

This chapter has examined educator roles in early childhood mathematics education. Educators are encouraged to see themselves as enculturating children to a culture of mathematics, through a range of pedagogical and communication strategies. Importantly, play is prioritised as a meaningful socio-cultural context in which to explore mathematical ideas.

The specific topics we covered in this chapter were:

- educators as mathematical enculturators
- playful pedagogical approaches
- child-instigated and educator-instigated experiences
- indirect and direct teaching
- intentional teaching
- responsive strategies
- listening and observing
- responding
- questioning
- collaborating
- modelling
- scaffolding
- communication roles
- the role of reflective practice
- shared contexts and shared responsibilities.

FOR FURTHER REFLECTION

A responsive mathematics education program requires educators to think carefully about the roles they enact and the strategies they use. To help you plan responsive mathematics programs, consider the following:

1 Why is it important to use different pedagogical approaches at different times?
2 How can different approaches allow children to access different mathematical understandings?
3 What type of educator role do you feel most comfortable with, and why?
4 How does your preferred educator role impact the nature of the mathematical learning experiences you provide for children?
5 How can you embed both reflection in action and reflection on action in your mathematics education practices?

FURTHER READING

Macmillan, A. (2009). The role of language in learning. In *Numeracy in early childhood: Shared contexts for teaching and learning* (pp. 48–65). South Melbourne, VIC: Oxford University Press.

Macmillan, A. (2009). Responsive and restrictive teaching practices. In *Numeracy in early childhood: Shared contexts for teaching and learning* (pp. 150–67). South Melbourne, VIC: Oxford University Press.

Siemon, D., Warren, E., Beswick, K., Faragher, R., Miller, J., Horne, M., Jazby, D., & Breed, M. (2021). *Teaching mathematics: Foundations to middle years*. South Melbourne, VIC: Oxford University Press.

06 Involving Families

Learn more
Go online to see a video from Paige Lee explaining the key messages of this chapter.

CHAPTER OVERVIEW

This chapter focuses on the critical role of families in early childhood mathematics education. The chapter presents a number of strategies for helping families to be involved in the early childhood mathematics education program so that their knowledge about their children's mathematics, and their children's mathematical learning at home, can be valued and built upon in the educational setting.

In this chapter, you will learn about the following topics:

- communicating with families
- building partnerships with families
- *Let's Count*
- family gatherings
- learning stories
- culture and language.

KEY TERMS

- Culturally responsive teaching practices
- Family gatherings
- Learning stories
- Relationships

AMY MACDONALD

Introduction

Families are children's first, and most important, teachers of mathematics. Through everyday experiences with their families, children learn to use mathematics in meaningful ways—be this helping dad to sort the washing, helping mum to work out how much lawn seed to buy, or negotiating with their siblings about whose turn it is on the Xbox. Recent research by Lee (2022) has shown how babies and toddlers engage in everyday mathematical interactions with their families, such as moving 'back and forth' while being pushed on a swing, counting the beat of music and their steps while dancing, or getting 'one bowl, one spoon and one bib' at breakfast time.

However, it may be the case that families are not recognised as part of children's mathematics education. This may be because educators are unaware of the powerful mathematical learning taking place in children's home lives; or, indeed, families themselves may not be aware of the role they are playing. Consider, for example, Figure 6.1: Together, this father and daughter are exploring ideas associated with space, location, and distance as they cross the creek. Do they consider this exploration as mathematical? Would the child's early childhood educators be aware of this mathematical exploration in everyday family activity? How might such an experience be jointly constructed as mathematical? One way of helping families to recognise the important role they play is by actively building partnerships between families and educators to support children's mathematics education. The 'coming together' of families and educators is important for children's mathematical learning, because the best outcomes for children occur when those involved in their lives work together (Goff, 2016).

FIGURE 6.1 Everyday mathematical learning in families

(Source: Paige Lee)

Children's parents (or caregivers) want to be involved in their education—they enjoy reading to their children, helping them learn the alphabet, and teaching them how to count (Coates & Thompson, 1999). However, many parents don't know how to go beyond these activities, especially in mathematics (Coates & Thompson, 1999):

> Parents may be unsure where to begin; some may be hesitant because of their own negative experiences with mathematics and school. But no matter their background, they are eager for opportunities to become involved that are respectful, supportive, and non-threatening (p. 211).

This chapter explores the ways in which families can be involved in children's mathematics education, both at home and within the early childhood service or school. The key premise underpinning this chapter is that of **relationships**. Relationships are based on mutual trust and respect. Relationships between educators and families help families to feel that their knowledge about their children is valued, and to know that their everyday family life is rich with opportunities to learn about mathematics. These

Relationships are connections between people that are based on mutual trust and respect.

relationships also help educators in schools and early childhood services to plan mathematics education programs that draw upon the knowledge they gain from families and help children make connections between the mathematics they explore at home and the mathematics they are learning about at the centre or school. This chapter presents specific strategies for building such relationships and involving families in their children's mathematics education.

Opinion Piece 6.1

ASSOCIATE PROFESSOR WENDY GOFF

I believe that when the adults in the lives of young children come together to support mathematical learning, a variety of opportunities emerge. Coming together provides an avenue for questioning and learning about mathematics, and also an avenue for sharing and learning about the mathematical understandings that children are developing. It positions adults in ways that enable them to better support and facilitate the continuity of mathematics across different contexts (home and school), and as a result, improves mathematical learning outcomes for children. Families are children's first teachers of mathematics. When they come together with educators to support the mathematical learning of children, they offer an insight into the mathematical histories of children that might otherwise remain elusive. Likewise, when educators come together to work with families, they offer an insight into the mathematical learning that children are developing. I believe that the adults in the lives of young children contribute enormously to the mathematical opportunities and experiences that are afforded to children. When they come together to support the mathematical learning of children, the opportunities for learning within different contexts are not only recognised, but they are also enhanced.

Wendy Goff is Associate Professor of School Partnerships and Early Years Mathematics Education at Swinburne University, Melbourne, Australia. She has worked extensively with schools to develop a variety of intervention and support programs targeting the social welfare needs of children and families. Her research focuses on adult relationships and how they might impact on the learning and development of children. Throughout her career she has worked as a social welfare worker, a pre-school teacher, and a primary school teacher. Wendy is particularly interested in the mathematical learning of children, and her recent work involved the implementation of an intervention focusing on adults noticing and supporting the mathematical understandings of children making the transition to school.

Learn more
Go online to hear Associate Professor Wendy Goff read this Opinion Piece.

Communicating with families

A first step in involving families in their children's mathematics education is communicating with families about their children's mathematical learning and development. Communication is key to helping families recognise their children's mathematical learning in their educational contexts. Communication helps to forge connections between children's mathematical learning in their early childhood education setting and the opportunities for exploring mathematics at home. Mathematics can easily be the focus of everyday communication strategies. For instance, Figure 6.2 shows how a focus on mathematics can be embedded within a childcare centre's daily report sent home to families.

FIGURE 6.2 Communicating with families about children's mathematics

Toddler Daily Reflection

March 9 th 2017

Another wonderful day in the Toddler room!

Today we were again participating in the indoor/outdoor program. But today most of us chose to play outside as our Educators set up some new and exciting things up outside for us to engage and interact with.

Again the sandpit was the place to be! Penny set up some more filling and pouring activities. But expanded it so we had more concepts going on. (Wet and dry.) (Light and heavy). Our Educators were right by our side to guide us with developing our language and knowledge behind this activity.

The swing was again a very popular place. We are able to sit down together and enjoy the soothing swinging motion and talk to each other about it. "Weee" We would say or "This is fun." said Bronte.

Our soft fall play area was move around today. Lots of new areas for us to climb and explore so we can develop or gross motor skills as well as our hand- eye coordination. Our favourite today was the slide.

Penny set up some different kinds of blocks for us too. As a team, some of us were able to stack them up into a very tall tower! We had to get Penny to help us reach the top as we got to the end. Once we finished. Penny let go of the tower and it came tumbling down. Everyone screamed and laughed. It was a very exciting time for us all.

Kate set up a lovely drawing space for us today. We were able to come and go as we pleased. As she kept our drawings for us to come back to. This space was also a good opportunity for us have simply sit down and have a drink of water with our peers and sit in the shade. This play space was very popular and will continue on.

Inside, Kerrie had lots of interesting conversations with a few of us. Andie, Pia and Lindsay were all busy at the tables talking about their puzzles. Discussions of shapes, pictures and colours were occurring. Kerrie would ask us lots of questions and we were happy and confident to answer them. From this activity, Kerrie wants to incorporate more colour play into our program, so we are going to keep our eye out for some new activities within the room.

TODAY'S MENU

Morning Tea: Fresh fruit platter

Lunch: Assorted sandwiches

Afternoon Tea: Rice cakes with Vegemite and Cream cheese Thank you Toddlers and Families for another beautiful day in the Toddler room.

(Source: Penny Baker/Amy MacDonald)

How confident do you feel about communicating with families about their children's mathematical learning? What strategies would you feel comfortable using?

Building partnerships with families

'When strong, collaborative partnerships have been built among families and early childhood educators, there is great potential for the educator to influence the families' thinking around their children's mathematics' (Perry & Gervasoni, 2012, p. 21). The importance of partnerships with families is recognised within the *Early Years Learning Framework for Australia* (EYLF; DET, 2019):

> Learning outcomes are most likely to be achieved when early childhood educators work in partnership with families. Educators recognise that families are children's first and most influential teachers. They create a welcoming environment where all children and families are respected and actively encouraged to collaborate with educators about curriculum decisions in order to ensure that learning experiences are meaningful (p. 13).

Each person involved in a partnership has something unique to contribute (Goff, 2016). 'The most productive way forward is to focus on what each participant—parent, teacher, community member—can bring to the partnership that will make the best use of their diverse expertise, backgrounds, and interests in supporting the child's numeracy learning' (Goos, Lowrie, & Jolly, 2007, p. 13). However, some family members will be reluctant to get involved with their children's mathematics education, and early childhood educators might have to develop specific strategies to support such involvement. In order to form partnerships, 'educators need to engage with families and communities in ways that are relevant, meaningful and culturally appropriate' (Perry & Gervasoni, 2012, p. 5). Gervasoni and Perry (2012) suggest that possible strategies for engaging parents and family members might include initiating conversations when children are collected to go home or inviting parents and family members to have a discussion over a cup of coffee. Working in community settings can be an effective way of reaching parents who have felt alienated from schools and other formal educational settings (Coates & Thompson, 1999). While families may be reluctant to participate in mathematics education programs that are school based, they are often more comfortable exploring mathematics and interacting with educators in community settings such as parks, libraries, and community centres.

What strategies can you think of to engage parents and other family members in children's mathematical learning and development?

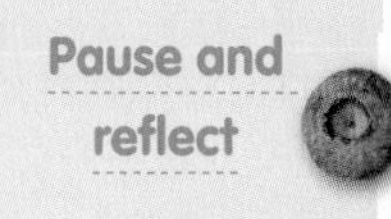

Let's Count

Let's Count is an early mathematics program designed by The Smith Family[1] and researchers Bob Perry and Ann Gervasoni as a means of assisting parents and other family members to help their young children play with, investigate, and learn powerful mathematical ideas. *Let's Count* involves early childhood educators in the role of mentors to the parents and family members of the children in their settings, providing assistance in noticing, exploring, and talking about mathematics in everyday life. *Let's Count* was developed with the aim of facilitating environments in which young children are stimulated to develop mathematical knowledge: It seeks to enable, according to Perry and Gervasoni (2012):

- partnerships among early childhood educators and families
- play and investigation for all
- recognition of all as potentially powerful mathematicians
- realisation that mathematical learning can be fun for all when it is undertaken in a relevant and meaningful context
- mentoring and advising of families by early childhood educators
- meaningful documentation of learning
- strong links to the theoretical and practical bases of the EYLF.

The *Let's Count* program requires early childhood educators to engage with two key pedagogical approaches: family gatherings and learning stories.

FIGURE 6.3 Face-to-face family gathering in the community

(Source: Vicki Olds)

1 The Smith Family is an Australian children's charity focused on 'helping disadvantaged Australian children to get the most out of their education, so they can create better futures for themselves' (www.thesmithfamily.com.au).

FAMILY GATHERINGS

The *Let's Count* program encourages educators to implement family gatherings with the children, parents, and family members in their setting. As MacDonald (2015) explains, **family gatherings** are essentially workshops designed to allow early childhood educators to have conversations about mathematics with parents, and to assist parents to explore mathematics with their children. Family gatherings are an opportunity for educators to work with families to assist them in recognising the opportunities for children's mathematical development in everyday family life. They enable educators to learn about, and appreciate, the unique resources and capacities of each family.

Family gatherings are workshops designed to assist educators and families to explore mathematics together.

Family gatherings may take any number of forms. Educators are encouraged to think about what might work best for the families in their service or school, and to think creatively about how they might 'gather' families around the topic of children's mathematical development. Family gatherings may differ significantly in size and setting; they may be undertaken in person or online. Indeed, they might look like any of the following:

- large group workshops
- sessions conducted with a small selection of families
- face-to-face gatherings, such as meetings at the early childhood centre or at a community site such as a park or playground (see Figure 6.3)
- virtual gatherings—for example, via email or closed Facebook groups (see Figure 6.4)
- one-off workshops or meetings
- ongoing communication and collaboration.

In reflecting upon the family gatherings in which they have been involved (see MacDonald, 2015b), educators have talked about how the *Let's Count* program helped them to develop their ability to act as mentors to parents and support the exploration of mathematics in home environments. Educators addressed the importance of the mathematical learning opportunities provided by parents, and they reported increases in parents' confidence as a result of the family gatherings. Educators also discussed how the family gatherings fostered relationships and partnerships between themselves and families. In particular, they commented on how the sense of partnership that developed between themselves and the families created different understandings of each other's roles and fostered mutual knowledge and appreciation.

FIGURE 6.4 Virtual family gathering via Facebook

(Source: Sarah Barcala)

LEARNING STORIES

Learning stories are qualitative snapshots of learning recorded as written narratives, often with accompanying photographs.

A second pedagogical approach employed in *Let's Count* is the writing of mathematical learning stories. **Learning stories** are qualitative snapshots of learning recorded as written narratives, often with accompanying photographs (Carr, 2001). Educators are encouraged to include three key features in their mathematics-focused learning stories:

1. description of the context and what happened
2. analysis of the child's mathematical learning
3. suggestions for how this learning might be further developed (with a focus on what families can do).

Learn more
Go online to hear Jody Rumble read this Educator reflection.

Educator reflection

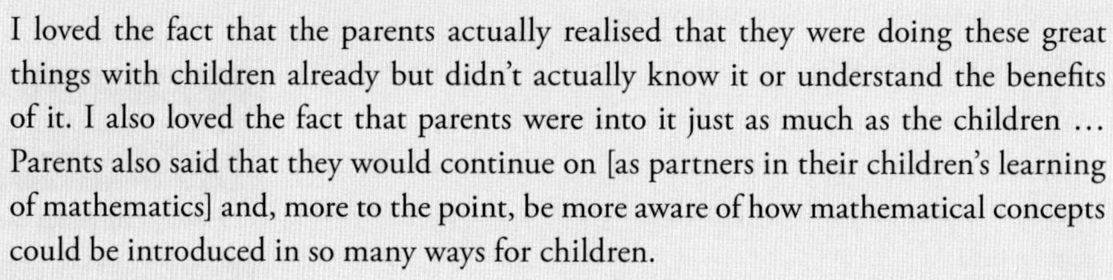

I loved the fact that the parents actually realised that they were doing these great things with children already but didn't actually know it or understand the benefits of it. I also loved the fact that parents were into it just as much as the children … Parents also said that they would continue on [as partners in their children's learning of mathematics] and, more to the point, be more aware of how mathematical concepts could be introduced in so many ways for children.

(Source: Jody Rumble)

Learn more
Go online to hear this Educator reflection being read.

Educator reflection

Through working on such projects with children and families as equal partners we are enabled to share and celebrate children's learning. The family I worked with were clearly proud of the child's numeracy understanding and thinking. The child was seen as competent by all and her family expressed an intention to further extend on her numeracy learning in their everyday lives.

(Source: Sarah Barcala)

The intent of this approach is to assist early childhood educators in noticing, exploring, and talking about children's mathematical learning. In this way, educators can attend more closely to the potential for mathematical development in children's play and investigation, and also hone their skills in communicating children's mathematical learning to others (MacDonald, 2015b). In particular, learning stories can be a communication tool for discussing children's mathematical learning with families, and provide a means of offering support and advice to families as to how they might explore mathematics at home with their child (MacDonald, 2015b).

An example of a learning story produced by a *Let's Count* educator is presented in Figure 6.5. More information about writing mathematics-focused learning stories is provided in Chapter 11.

FIGURE 6.5 Example of a mathematical learning story

<table>
<tr><th>CHILD'S NAME: E
AGE: 4 YEARS OLD
EDUCATOR: STEPHANIE</th><th>DATE: 23/4/13</th></tr>
<tr><th>EYLF OUTCOMES</th><th>LEARNING STORY: BLOCK PLAY</th></tr>
<tr><td>Outcome 1: Children have a strong sense of identity.
a Children feel safe, secure and supported
b Children develop their emerging autonomy, inter-dependence, resilience and sense of agency
c Children develop knowledgeable and confident self-identities
d Children learn to interact in relation to others with care, empathy and respect.
Outcome 2: Children are connected with, and contribute to, their world.
a Children develop a sense of belonging to groups and communities and an understanding of the reciprocal rights and responsibilities necessary for active civic participation
b Children respond to diversity with respect
c Children become aware of fairness
d Children become socially responsible and show respect for the environment.
Outcome 3: Children have a strong sense of wellbeing.
a Children become strong in their social, emotional and spiritual wellbeing
b Children take increasing responsibility for their own health and physical wellbeing.
Outcome 4: Children are confident and involved learners.
a Children develop dispositions for learning such as curiosity, cooperation, confidence, creativity, commitment, enthusiasm, persistence, imagination and reflexivity
b Children develop a range of skills and processes such as problem solving, inquiry, experimentation, hypothesising, researching and investigating
c Children transfer and adapt what they have learnt from one context to another
d Children resource their own learning through connecting with people, place, technologies and natural and processed materials.
Outcome 5: Children are effective communicators.
a Children interact verbally and non-verbally with others for a range of purposes
b Children engage with a range of texts and get meaning from these texts
c Children express ideas and make meaning using a range of media
d Children begin to understand how symbols and pattern systems work
e Children use information and communication technologies to access information, investigate ideas and represent their thinking.</td><td>What happened:
E was sitting alone when she had set up her blocks onto the table. While there I noticed she was talking to herself while building. I tried to stand in closer to see what she was saying when another child joined her. E appeared to be making a little town and talking herself through the steps: 'we need 1 block for the house, 1 block for the roof, we also need more houses so we will need more blocks for this'. E appeared to be devising a plan before she started her work. She knew exactly what she wanted to build but when the other child came into the area and saw what she was building they started to build together. E led this play and showed great verbal direction and communication with her peer. E was able to name the blocks and give each a number. She didn't go higher than 3 at a time, and her blocks were systematically lined up next to each other along the table.
Mathematical learning that took place:
E developed this play situation from her own ideas and sense of agency and autonomy (Outcome 1b). Once again she shows great understanding of persistence, enthusiasm, commitment, and imagination (Outcome 4a). E's problem solving and hypothesising skills are once again key in her play skills and the high level of understanding she has with these two very abstract terms (Outcome 4b). She is able to count the blocks and predict which ones will make the houses. She demonstrated the concept of understanding and naming numbers and that by placing each on top of the other she identified the outcome of this process (it makes a house). E can interpret and perceive different levels within her building. E shows sound knowledge in shapes and patterns and she does this by turning a triangular block into a roof or a square block into a base for a house (Outcome 5d). E's spatial and measurement awareness when constructing the building out of the blocks shows she is experimenting with height, width, length, and three-dimensional equipment.
What next:
Add pictures of multi-storey houses, so she can expand her awareness of shapes, patterns, and levels. This will then encourage and extend on her counting abilities, shape naming, and recognition skills. E's parents could take her to the city to see some high-rise buildings or show her pictures on their computer or in magazines to encourage her imagination.</td></tr>
</table>

(Source: Stephanie Hill)

Culture and language

Culturally responsive teaching practices recognise and respond to the diverse cultural and language backgrounds of children and families.

Culturally responsive teaching practices are important for helping families to engage with their children's mathematics education. Many families have diverse cultural and language backgrounds and may, as a result, use mathematics in ways that are culturally familiar but different to those of the educational setting. In particular, this is often a challenge faced by Indigenous children and families, whose discourses may differ from those of the early childhood service or school (Cairney, 2003). A growing body of research focuses on the value of recognising or incorporating culturally linked knowledge and practices into mathematics teaching and learning (Vale, Atweh, Averill, & Skourdoumbis, 2016). Partnerships with families are key to effective culturally responsive mathematics education practices (Vale et al., 2016), as families can provide insight into differing cultural practices and languages of mathematics.

Language is critical in early childhood mathematics education. Mathematical language, in itself, presents challenges for young children, as the meanings of words used in relation to mathematics often differ from their meanings in everyday language (Perry & Gervasoni, 2012). This may be all the more challenging for children and families for whom English is not their first language (Perry & Dockett, 2008):

> Children need sufficient language to allow them to understand their peers and their teachers as explanations are presented, and to allow them to give their own explanations. This has particular ramifications for those children for whom the language of instruction is not their first language ... This situation is often exacerbated in the development of mathematical ideas because of its specialised vocabulary and its use of 'common' words to have specialised meanings (p. 93).

Early childhood educators can help to foster links between children's home language and the language of mathematics education by involving families in the process. Educators can encourage parents to talk about mathematics in different languages, to teach their children to count in their home language, and to play the maths games they played as children (Gervasoni & Perry, 2012). Educators can help create a bridge between Aboriginal English and Standard Australian English for young Indigenous children as they grapple with new language, concepts, and vocabulary for mathematics (Warren, Young, & De Vries, 2008). Making connections between children's home language and the language of instruction can help children to feel more confident about their mathematical knowledge and learning. Perry and Gervasoni (2012) give the following example:

> Imagine the joy and cultural connection felt by a 4-year-old boy who is able to count in Standard Australian English, and in his home language(s), and also in the languages of his friends. If such joy is celebrated by the child's family and early childhood educators, imagine the esteem this will build in the child and the learning that other children will experience (p. 18).

How might you find out about the cultural backgrounds of the children and families in your setting? How could this cultural knowledge be embedded in the mathematics education program?

Educator reflection

We have a few families at our centre that are bilingual. As a way of connecting with these families we ask parents to provide us with key words so we can learn some of their language. We use maths as a way of transitioning throughout the day, usually through activities like counting, number recognition, shapes, etc. We asked one family to teach us to count to twelve in French; they happily wrote the numbers in French and gave us lessons so we could share their culture and their knowledge with all the children.

(Source: Michelle Gunter)

Learn more
Go online to hear this Educator reflection being read.

Chapter summary

Families are children's first and most important teachers of mathematics, and opportunities to learn about mathematics exist in the everyday lives of all children and families. This chapter has outlined a range of strategies for supporting the involvement of families in children's mathematics education. Building partnerships with families helps families to feel that they are an important part of their children's mathematics education. Such partnerships also help educators to learn more about the mathematics in children's everyday lives, and to embed this knowledge in the mathematics education program in meaningful ways.

The specific topics we covered in this chapter were:

- communicating with families
- building partnerships with families
- *Let's Count*
- family gatherings
- learning stories
- culture and language.

FOR FURTHER REFLECTION

Relationships with families are built on mutual trust and respect. Consider the following:

1 How can you help families to feel that they are trusted partners in their child's mathematics education?
2 How can you help families to see you as a trusted partner in their child's mathematics education?
3 What strategies could you implement in your mathematics education program to invite input from families?
4 How can you communicate the value of family input to families?

FURTHER READING

Coates, G.D., & Thompson, V. (1999). Involving parents of four- and five-year-olds in their children's mathematics education: The FAMILY MATH experience. In J.V. Copley (ed.), *Mathematics in the early years* (pp. 205–14). Reston, VA: National Council of Teachers of Mathematics.

MacDonald, A. (2015b). *Let's Count*: Early childhood educators and families working in partnership to support young children's transitions in mathematics education. In B. Perry, A. MacDonald, & A. Gervasoni (eds), *Mathematics and transition to school: International perspectives* (pp. 85–102). Singapore: Springer.

07 Strengths Approaches

Learn more
Go online to see a video from Paige Lee explaining the key messages of this chapter.

CHAPTER OVERVIEW

This chapter will explore the idea of 'strengths approaches', a concept more common in the social work sector, but one that is growing in recognition in early childhood education. A general explanation of strengths approaches will be provided, along with discussion of how this approach might inform early childhood mathematics education.

In this chapter, you will learn about the following topics:

» a social justice perspective
» strengths-based practice
» a strengths view vs a deficits view
» the strengths approach
» the column approach
» strengths-based planning in mathematics education
» using the column approach to articulate a goal
» strengths-based learning plans.

KEY TERMS

» Column approach
» Competency cycle
» Complex circumstances
» Deficit cycle
» Social justice
» Socio-economic status (SES)
» Strengths
» Strengths approach
» Strengths-based learning plans
» Strengths-based practice

AMY MACDONALD

Introduction

Some years ago, I was marking a cohort of pre-service teachers' mathematics education essays when I came across a student's statement worded as follows: 'poor kids can't do normal maths'. I was shocked at such a statement. How could the student have formed such a deficits view of particular children's mathematical abilities? Who are these 'poor kids', and what is 'normal maths' anyway? But after a few years of reflection, I can see how such a view might have been constructed as a result of a pervasive discourse in the mathematics education literature—that is, the perception that children who experience complex circumstances are disadvantaged in mathematics education.

Socio-economic status (SES)
is a measure of an individual or family's economic or social position in relation to others.

There is no disputing the fact that there is a large body of literature that supports this viewpoint. Research consistently shows that children from different **socio-economic status (SES)** backgrounds enter school with very different skills, with children from low-SES backgrounds performing lower than children from high-SES backgrounds in assessments of their school-entry mathematics skills (Carmichael, MacDonald, & McFarland-Piazza, 2014; Gould, 2014). Furthermore, research indicates that this performance gap exists not only when children start school, but as they continue through their schooling (Gould, 2014). This sort of data results in a perception that children who experience complex circumstances lack the necessary resources to succeed in mathematics education.

However, in this chapter, I present an alternative discourse: that all children, regardless of their family circumstances, possess unique resources that can be used effectively to allow children to thrive in mathematics. This chapter introduces a *strengths* framework for approaches to early childhood mathematics education.

Opinion Piece 7.1

Learn more
Go online to hear Associate Professor Angela Fenton read this Opinion Piece.

ASSOCIATE PROFESSOR ANGELA FENTON

I first came across strengths approaches as an early childhood educator working with children and families in cross-disciplinary settings. The social workers and therapists I was working with were using this approach and gaining excellent results with 'hard to reach' children and supporting families with multiple complex issues. I had not heard of the approach in my education studies and yet it seemed so relevant and practical—I wanted to learn more. I did, and started to practise the approach myself. Later I began to research how the approach may engage children with many curriculum areas—including early childhood mathematics.

I had my own personal experiences from childhood that were based on maths teaching using a deficits model. My memories were crowded with the maths concepts that I had failed to comprehend, questionable teaching techniques, frustration, and a lack of confidence in maths that followed me into adulthood. Yet, when I see a strengths approach to mathematics in action I see the opposite occurring. I see confident young learners—recognising and building on their strengths, problem solving and excited about their achievements. I also see teachers who guide and support children, quietly scaffolding and celebrating with them their mathematical development.

Angela Fenton is an Associate Professor of Early Childhood Education at Charles Sturt University, Albury–Wodonga, Australia. Angela's research centres on early childhood development and wellbeing, child protection, and early childhood teacher education. She is examining the use of a strengths approach in pre-service teaching to prepare students for working with children who may be at risk or experiencing child abuse. Angela's most recent research examines the significant opportunities in using a strengths approach in early childhood education to enhance children's ability to thrive in all areas of development. Angela worked in early childhood classrooms and services for over twenty years as a teacher, director, and training project officer. Angela has worked in inclusion roles with children with disabilities and was the state manager for the Indigenous Children's Services Unit in Queensland. She has taught in the United Kingdom and in Australia.

A social justice perspective

In our work with young children, it is certain that we will encounter children and families who experience **complex circumstances**. Families with complex circumstances are those experiencing various challenges such as low income or unemployment, cultural diversity, physical or mental health issues, violence, or trauma, to name just a few. Families who experience these complexities are less likely than other families to be engaged in the educational setting and have positive relationships with educators (Fenton & McFarland-Piazza, 2014).

Complex circumstances refers to social, financial, cultural, or health challenges experienced by families.

Consider the children and families you know, or with whom you have worked. What challenges might they be experiencing as a family? How might these challenges affect their experiences in educational settings, and, in particular, their relationships with educators?

It is sometimes the case that a child's family circumstances have a significant impact upon their experiences of mathematics education. As Atweh, Vale, and Walshaw (2012) explain, children

> experience the education of mathematics differently, based on their learning opportunities and achievements that depend on the social context of their families … Often such 'background' factors are associated with disadvantage, marginalisation, disengagement, and exclusion from the study of mathematics (p. 39).

Educator reflection

I learnt maths in primary school at a time when maths books were used only if your parents could afford them. If like my family they couldn't, then you were left behind and listening in a class without assistance, a textbook, or any help from the teacher. Needless to say I got 'modified' maths results. When I began working in childcare (all those years ago) I made a conscious decision that all children in my care would have the same opportunities as each other to learn. I make learning fun and choose a variety of ways to implement the learning suitable to types of learners.

(Source: Lynnette Hartley)

Learn more
Go online to hear this Educator reflection being read.

Lynnette's reflection makes a clear point about ensuring that all children, regardless of their personal circumstances, have the same opportunities to learn about mathematics. In view of this, a social justice perspective of mathematics education seeks to address the potential inequities that children might experience on the basis of their differing family circumstances. Broadly speaking, **social justice** is about valuing diversity and providing equal opportunities. In mathematics education, this means providing learning experiences that allow *all* children, regardless of the challenges they face in their lives, the opportunity to *thrive mathematically*.

Social justice
is about valuing diversity and providing equal opportunities.

Pause and reflect

What does the term 'social justice' mean to you? What are some of the social justice issues that you might encounter in early childhood mathematics education?

Strengths-based practice

One such way of providing opportunities for children to thrive is by implementing strengths-based practices in mathematics education. Current early years curriculum and policy documents such as the *Early Years Learning Framework for Australia* (EYLF; DET, 2019), the *Framework for School Age Care* (DEEWR, 2011), and the *National Quality Standard* (NQS) (Australian Children's Education and Care Quality Authority, 2013) advocate the use of strengths-based practices in early childhood settings. For example, the EYLF contains the following strengths-based statements:

> Educators are responsive to all children's strengths, abilities and interests. They value and build on children's *strengths*, skills and knowledge to ensure their motivation and engagement in learning (p. 16).
>
> Knowledge of individual children, their *strengths* and capabilities will guide educators' professional judgement to ensure all children are engaging in a range of experiences across all the Learning Outcomes in ways that optimise their learning (p. 22).
>
> When early childhood educators respect the diversity of families and communities, and the aspirations they hold for children, they are able to foster children's motivation to learn and reinforce their sense of themselves as competent learners. They make curriculum decisions that uphold all children's rights to have their cultures, identities, abilities and *strengths* acknowledged and valued, and respond to the complexity of children's and families' lives (p. 14; italics added for emphasis).

But what, exactly, is meant by the term 'strengths'?

Strengths can be defined as people's intellectual, physical, and interpersonal skills, capacities, interests, and motivations (Mallucio, 1981, cited in McCashen, 2005). A person's strengths can also include the resources in their environment, such as family, friends, neighbours, colleagues, material resources, and so on (McCashen, 2005).

Strengths
can be defined as people's intellectual, physical, and interpersonal skills, capacities, interests, and motivations.

With this definition in mind, **strengths-based practice** in early childhood education can thus be considered as educational practice that recognises, and utilises, children's strengths. It emphasises the notion of starting from what children *can* do, rather than what they *cannot*.

Strengths-based practice is educational practice that recognises and utilises children's strengths.

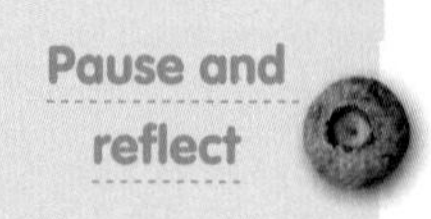

How would you define strengths-based practice in early childhood mathematics education?

Educator reflection

The majority of my families walk to kindergarten so I engage in everyday conversations with the children and families. For example, I ask them such questions as 'How long did it take you to walk today?' 'Did you have to wait for long at the lights?' and 'Were there many people and cars on your way?' I have been sending home books for the children and families to read together that involve mathematical concepts, and I have put up picture displays for the parents to see how their children engage in mathematics during play (blocks, cooking, puzzles, drawing). This has been a great way to involve parent input and create discussions around mathematical learning with families in my centre.

(Source: Patricia Woodford)

Learn more
Go online to hear this Educator reflection being read.

A strengths view vs a deficits view

As outlined in the Introduction to this chapter, mathematics education literature often presents a deficits view of children's mathematical abilities. Children are often characterised by what they *cannot* do, rather than by what they *can* do. This is particularly the case for children from family circumstances that may be considered 'complex' or 'disadvantaged'.

A **deficit cycle** focuses on negative expectations and results in negative experiences.

A deficits view of children's mathematical ability can have self-reinforcing negative effects, whereby negative expectations of children's learning abilities result in negative experiences in mathematics education. This can be described as a **deficit cycle**—an inadvertent process of disempowerment that cannot reverse the negative expectations, experiences, and behaviours that it establishes from the outset (McCashen, 2005) (see Figure 7.1). In contrast, a focus on strengths emphasises children's competencies and resources, which can be utilised to bring about positive change in their mathematics education experiences. This can be described as a **competency cycle**—a process of creating positive expectations and opening the way for the development of new competencies (McCashen, 2005) (see Figure 7.2).

A **competency cycle** focuses on strengths and positive expectations and results in the development of new competencies.

Pause and reflect

How is a deficit cycle different from a competency cycle? How can a competency cycle be effected in an early childhood mathematics education program?

FIGURE 7.1 A deficit cycle

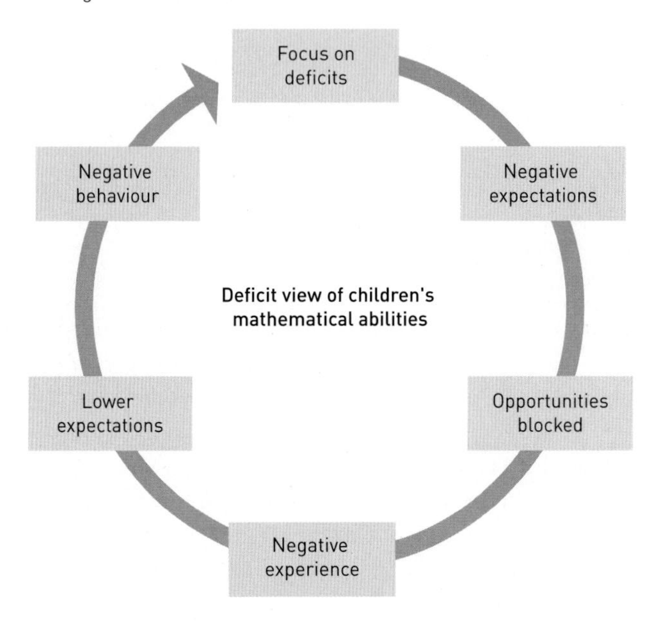

(Adapted from McCashen, 2005)

FIGURE 7.2 A competency cycle

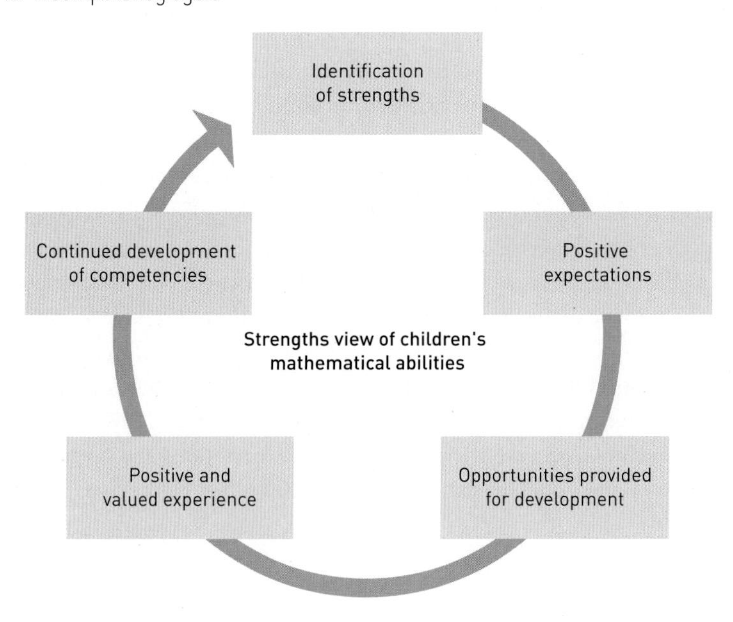

(Adapted from McCashen, 2005)

The strengths approach

In Australia, the social service organisation St Luke's, based in Bendigo, Victoria, pioneered the use of strengths-based practices in their work with families experiencing complex circumstances. This resulted in the development of the **strengths approach** (McCashen, 2005)—a way of working with children and families that encourages the identification of resources and the use of challenges, as they occur, to create resilience and aptitude when working with issues (Fenton, MacDonald, & McFarland-Piazza, 2016).

The **strengths approach** is a way of working with children and families to help build their resilience.

The strengths approach, as outlined in McCashen (2005, pp. 47–8), is based on six key stages of reflection, planning, and action:

1 listening to children's and families' stories, exploring the context, and identifying the core issues
2 developing a picture of the future and setting goals
3 identifying and highlighting strengths and exceptions to problems
4 identifying additional resources that complement people's strengths and goals
5 mobilising strengths and resources through a plan of action
6 reviewing and evaluating progress and change.

The first five stages of the strengths approach are usually presented in a five-column framework that can be used as a guide to the practical application of the approach. McCashen (2005) describes this as the 'column approach'.

THE COLUMN APPROACH

The **column approach** is a tool for implementing the strengths approach in a variety of practical circumstances (Fenton et al., 2016). This is often referred to as a 'strengths-based process' (McCashen, 2005). The column approach provides a structured framework for thinking through the first five stages of the strengths approach, and for applying them in practical ways. The framework for implementing a strengths-based process using the column approach is shown in Table 7.1.

The **column approach** is a tool for implementing the first five stages of the strengths approach.

TABLE 7.1 The column approach

STORIES AND ISSUES	THE PICTURE OF THE FUTURE	STRENGTHS AND EXCEPTIONS	OTHER RESOURCES	PLANS AND STEPS
Ask questions that invite children and families to share their stories and enable them to clarify the issues, for example: What's happening? How do you feel about this? How long has this been a problem? How is it affecting you and others?	Ask questions that help children and families explore their aspirations, interests, and goals, for example: What do you want to be happening instead? What will be different when the issues are addressed?	Ask questions that help children and families explore their strengths and the exceptions to the issues, for example: What strengths do you have that might be helpful? What do you do well? What's happening when the issues aren't around?	Ask questions that help children and families identify resources that might help them reach their goals, for example: Who else might be able to help? What other skills or resources might be useful?	Ask questions that enable children and families to specify concrete steps towards achievement of their goals, for example: What steps can be taken, given your picture of the future, and your strengths and resources? Who will do what? How? By when?

(Adapted from McCashen, 2005, p. 48)

CURRICULUM CONNECTIONS

Early Years Learning Framework

Outcome 1: Children have a strong sense of identity

Children develop knowledgeable and confident self-identities.

This is evident when children:

- Feel recognised and respected for who they are
- Celebrate and share their contributions and achievements with others.

Educators promote this learning when they:

- Ensure all children experience pride and confidence in their achievements
- Share children's successes with families
- Show respect for diversity, acknowledging the varying approaches of children, families, communities and cultures
- Demonstrate deep understanding of each child, their family and community contexts in planning for children's learning
- Build on the knowledge, languages and understandings that children bring
- Provide rich and diverse resources that reflect children's social worlds.

(DET, 2019, p. 26)

Strengths-based planning in mathematics education

Mathematics educators working with a strengths approach will focus on what mathematics children can do, as well as the opportunities and resources available to assist in the development of their strengths and capacities to meet identified learning goals (Collins & Fenton, 2021). As discussed in Chapter 6, families can help their children learn mathematics in their everyday lives, and early childhood educators can also offer opportunities for families to become involved in their children's mathematics (Fenton et al., 2016).

Learn more
Go online to hear Alexandra Roth read this Educator reflection.

Educator reflection

Last year I had a child in the baby room whose mother is Japanese. I recorded her counting 1–10 and used to play it quite often for the child. The child's face lit up every time I played it and I could see she loved having that connection with home, and hearing and understanding a language that was familiar.

(Source: Alexandra Roth)

A strengths approach can encourage collaborative partnerships between educators and families. On the other hand, a deficits approach—which focuses on children's and families' weaknesses—can often result in a stigma, such as shame about perceived inability to learn (Dockett et al., 2011). In this section, I will provide some examples of how McCashen's strengths approach and five-column framework can be adapted to mathematics education in order to draw upon the unique resources and capacities of *all* children and their families.

USING THE COLUMN APPROACH TO ARTICULATE A GOAL

The column approach is a useful strategy for educators to develop a narrative of their children's opportunities for learning in mathematics, identify the children's hopes and dreams, consider the children's strengths and mathematical capacities, identify resources that are available to them, and map out a way for them to move forward (Collins & Fenton, 2021). The five-column framework can be used to think through, and articulate a particular goal for, early childhood mathematics education. This approach encourages educators to focus on the strengths of those involved rather than their deficits, and to consider the resources that might be at their disposal to assist in achieving their goal. For instance, early childhood educator Jody Rumble was interested in developing a mathematics education program based around the mathematical learning opportunities in children's home environments. The column approach can be used to articulate this goal and identify the resources and steps for working towards this goal. An example is shown in Table 7.2.

TABLE 7.2 The column approach applied to mathematics education

STORIES AND ISSUES	THE PICTURE OF THE FUTURE	STRENGTHS AND EXCEPTIONS	OTHER RESOURCES	PLANS AND STEPS
The issue is how to incorporate mathematical learning into children's home environments.	The vision is to collaborate with families to recognise and support early mathematical learning in family contexts and to assist in transitions to school.	The educator brings these strengths: » early childhood developmental understandings » mathematical knowledge » communication skills » organisational skills. The parents bring these strengths: » deep knowledge of, and concern for, their children » safe and stimulating learning environments » willingness to put time and effort into the collaboration. The children bring these strengths: » existing mathematical understandings that can be built upon » a variety of interests.	The educator can provide some sample ideas and activities for exploring mathematics at home. The educator can also provide parents with some information on key mathematical concepts to look for and explore with the children.	The educator will work with parents to suggest ideas for incorporating mathematical learning and building on the strengths identified. A strengths-based learning plan will be developed for each child.

(Adapted from Fenton, MacDonald & McFarland-Piazza, 2016)

STRENGTHS-BASED LEARNING PLANS

Strengths-based learning plans are structured planning documents that scaffold planning on the basis of children's strengths and interests.

The column approach can be used to articulate a goal for mathematics education, and guide further planning to achieve this goal. **Strengths-based learning plans** can be used as structured planning documents for designing mathematical learning experiences that respond to children's strengths and interests. Building on the previous example, Jody was able to develop strengths-based learning plans that provide ideas for exploring mathematics at home, based on the children's interests and the families' strengths. Two examples of these plans are shown in Tables 7.3 and 7.4. This strengths-based approach to planning underpins the more detailed Learning Experience Plan approach that will be introduced in the next chapter.

Learn more
Go online to download a Learning plan template to create your own lesson plan.

TABLE 7.3 Jody's strengths-based learning plan for Child 1

STRENGTHS	IDEAS FOR LEARNING	MATHEMATICAL CONCEPTS
Interested in dinosaurs	Setting up several of his favourite dinosaurs in height order. Drawing his favourite dinosaurs in height order or heaviest to lightest, etc.	Sorting by size Size, area
Playing with his sister and dressing up as a superhero	Superhero game: Dress up as superhero and his sister has to hide, before the superhero finds her—he has to count backwards from 10 to zero before blast off. Hide some of the child's toys around the house and the superhero needs to find them.	Number order Counting on—how many has he found, how many left? Subtraction—how many more does he need to find?
Enjoys constructing with Lego	Building towers—discussion. Giving direction to where he should place some blocks. Sharing some blocks with his sister. Pattern making with Lego.	Height—how tall can you build it? Estimate—how many pieces do you think you have used? Position (describe)—on top, behind, in front of What would be fair—half to you and half to your sister? What comes next in a pattern?
Likes to climb trees	How long will it take you to get from the bottom to the top of the tree? How far is the tree from the house (or another object)?	Time, estimate Estimate, measurement—heel to toe, counting
Enjoys drawing	Who lives in our house? Drawing them. Game: Draw a square, draw a bigger square, draw a circle next to the smallest square, draw mummy on top of the big square, draw daddy on the inside of the circle, etc.	Size, heavy/light, weight, big/small Position, shapes, direction
Is interested in cooking	Cupcake making	Measurement, time, volume—half, full, quarter

(Adapted from Fenton et al., 2016)

TABLE 7.4 Jody's strengths-based learning plan for Child 2

STRENGTHS	IDEAS FOR LEARNING	MATHEMATICAL CONCEPTS
Enjoys riding her bike	How far can you ride in 30 seconds? Where can you ride? How far do you think it is from home to the end of the street? Mapping the bike ride out. How long will it take to ride that far? What is the probability that it won't rain while you are out riding?	Time, direction, position, area, counting, estimate
Likes playing with the dog	How far could you throw the ball for the dog? Walking the dog. Feed the dog half a cup of food. Can you put three cups of water in the dog's water bowl?	Estimate, distance, direction Measurement—how far can you walk? Direction, mapping prior to walking the dog Volume
Engages with computer games	Suggested age-appropriate mathematical websites: https://www.abc.net.au/abckids/games https://www.sesamestreet.org/games	Direction game Matching game Measurement—volume, more/less
Often spends time with doll play	Set up a doll's picnic—perhaps using four dolls. Get 16 pieces of food and have the child share the food between the dolls so they all get an equal amount of food. How is the food shared? For example, one by one, in groups of two or three at a time etc. until the food is all distributed.	Equal, sharing, sorting Sharing, classification
Is skilled at drawing	Who lives in our house? Drawing them. Game: Draw a square, draw a bigger square, draw a circle next to the smallest square, draw mummy on top of the big square, draw daddy on the inside of the circle, etc.	Size, heavy/light, weight, big/small Position, shapes, direction
Likes to watch and help with food preparation	Cooking cupcakes	Measurement, time, volume—half, full, quarter

(Adapted from Fenton et al., 2016)

Chapter summary

In this chapter I have introduced a strengths approach to early childhood mathematics education. This approach encourages educators to draw on the abilities, interests, resources, and capacities of children and their families when planning mathematical learning experiences. This chapter provides a strengths-based framework for mathematics education planning and pedagogy. I encourage you to revisit this framework as you continue to read this book, and to reflect on how this framework might influence all aspects of your mathematics education practice.

The specific topics we covered in this chapter were:

- a social justice perspective
- strengths-based practice
- a strengths view vs a deficits view
- the strengths approach
- the column approach
- strengths-based planning in mathematics education
- using the column approach to articulate a goal
- strengths-based learning plans.

FOR FURTHER REFLECTION

A social justice view of mathematics education requires a program that recognises, utilises, and builds upon each individual's strengths. To help you plan your mathematics programs, consider the following:

1 Why is it important for early childhood educators to implement a strengths-based process when planning for mathematics education?

2 What are the strengths of the children and families with whom you are working? How can these strengths be incorporated in the mathematics education program?

3 What are your own strengths? How can you make effective use of these strengths in your own mathematics education practice?

FURTHER READING

Fenton, A., MacDonald, A., & McFarland-Piazza, L. (2016). A Strengths Approach to supporting early mathematics learning in family contexts. *Australasian Journal of Early Childhood, 41*(1), 45–53.

McCashen, W. (2005). *The strengths approach: A strengths-based resource for sharing power and creating change.* Bendigo, VIC: St Luke's Innovative Resources.

08 Informal Opportunities to Learn

Learn more
Go online to see a video from Paige Lee explaining the key messages of this chapter.

CHAPTER OVERVIEW

This chapter will explore a range of everyday play contexts that provide opportunities for children to develop and use informal mathematics. These contexts are a part of most young children's everyday lives, and by noticing the mathematics within these play situations, educators can incorporate children's informal mathematics into the formal education program in meaningful ways.

In this chapter, you will learn about the following topics:

- mathematics in play
- outdoor play
- digital play
- techno-toys
- apps and games
- music and movement
- children's drawings
- storybooks.

KEY TERMS

- Contexts
- Digital play
- Digital technology
- Formal mathematics
- Informal mathematics
- Techno-toys

Introduction

Children engage in a variety of everyday activities that involve mathematics, and as a result develop a considerable body of informal mathematical knowledge in their early childhood years. Furthermore, children draw on this informal knowledge when engaging in more formal mathematics. It is quite often the case that the mathematics in informal activities is not immediately obvious; indeed, the mathematical learning opportunities in everyday play, drawing, or storybook reading often goes overlooked. However, these undertakings provide rich and meaningful **contexts**—interactions, activities, and settings—for children to access mathematical ideas.

Contexts consist of interactions, activities, and settings.

Geary (1994) draws attention to the ways in which mathematical activities are a part of young children's everyday lives, explaining that even two-year-olds regularly engage in mathematics-related activities:

> These activities include solitary episodes, such as counting toys or number of snacks, as well as social play. Parents and children frequently engage in number-related play, such as singing songs with number references (e.g., 'One, two, buckle my shoe') and counting toes and fingers, as well as activities that require an understanding of the uniqueness of individual digits (e.g., 'Please turn the TV to Channel 506') (pp. 11–12).

This chapter draws attention to the informal opportunities to learn about mathematics that exist in most children's everyday lives. **Informal mathematics** is that which is based on informal mathematical practices in everyday life. This is in contrast to **formal mathematics**, which uses formal and standardised practices and approaches to mathematics. However, the activities canvassed here can be integrated into formal mathematics education practices to provide meaningful links between the ways in which children explore mathematics at home and in their educational settings.

Informal mathematics uses informal mathematical practices in everyday life.

Formal mathematics uses formal and standardised practices and approaches to mathematics.

Mathematics in play

It is well established that play is essential to young children's learning. As stated in the *Early Years Learning Framework for Australia* (EYLF; DET, 2019, p. 10), play is a context for learning that:

- allows for the expression of personality and uniqueness
- enhances dispositions such as curiosity and creativity
- enables children to make connections between prior experiences and new learning
- assists children to develop relationships and concepts
- stimulates a sense of wellbeing.

Everyday play affords an opportunity for learning powerful mathematical ideas. For example, 'informal exposure to number concepts through activities such as game playing, cooking or completing jigsaw puzzles may help children to develop an awareness of numbers and quantities' (Skwarchuk & LeFevre, 2015, p. 113). Many mathematical ideas are also embedded in board games, card games, and technology-based games. Such materials and activities are a natural part of children's everyday life; special materials are not necessary to promote children's mathematical activities (Ginsburg, Inoue, & Seo, 1999).

What mathematics is used in the games you play with children? What questions might you ask to assist children to notice and explore the mathematics? How might you adapt the games so that the mathematics involved is easier, or more complex?

By way of example, Ginsburg et al. (1999) highlight the mathematics in construction play such as playing with wooden blocks. As Ginsburg et al. note, block play affords opportunities for spontaneous exploration of mathematical ideas such as length–distance relationships, reasoning about two-dimensional relationships among different components of block structures, and considering the lengths, widths, and angles of the structures. Indeed, block play is rich with opportunities for exploring shapes, spaces, patterns, and relationships (Figure 8.1).

FIGURE 8.1 Block play

(Source: Paige Lee)

Play contexts provide opportunities for children's spontaneous invention of their own mathematical strategies—that is, children use what they know to solve the mathematical problems with which they are faced. For example, children are often faced with problems involving sharing or distributing play materials, and they are required to come up with solutions that are fair for all players. Children also use their lived experiences to inform the play, and these often result in individual approaches to structuring the play, utilising the play materials, or solving the problems presented in the play.

Learn more
Go online to hear this Educator reflection being read.

Educator reflection

Growing up in an army family I got to see lots of parades and displays of military equipment. This to some degree had an influence on how I played at home with my toys. While I had a good collection of toy soldiers that I would play with, I would use all manner of objects to represent armies, assembling them into platoons, units, sections, divisions, all evenly set up in equal rows. Some examples included marbles, rocks, cigarette butts, bottle caps, tiles, Lego pieces, pegs, and sewing pins with coloured heads. There also had to be a leadership structure for these armies to follow with different ranks controlling the various-sized groupings. I used a lot of maths when I was playing. I'm a little surprised at how well I understood how this type of grouping worked and the hierarchy influenced them in terms of size.

(Source: Jason Goldsmith)

CURRICULUM CONNECTIONS

Early Years Learning Framework

Outcome 4: Children are confident and involved learners

Children resource their own learning through connecting with people, place, technologies and natural and processed materials.

This is evident when children:

- Explore ideas and theories using imagination, creativity and play.

Educators promote this learning when they:

- Think carefully about how children are grouped for play, considering possibilities for peer scaffolding.

(DET, 2019, p. 40)

Outdoor play

The outdoor play environment provides a rich resource for exploring mathematical concepts. Outdoor environments provide different ways of exploring spaces and relationships within those spaces. In particular, the nature of outdoor environments promotes physical play in which children can use their bodies to test out ideas and explore spaces. Consider, for example, the different spatial understandings being developed as children climb *up* ladders, slide *down* slippery-dips, crawl *through* tunnels, and jump *into* puddles. This embodied activity gives tangible meaning to spatial concepts, as children are able to position their bodies in spaces in different ways (Figure 8.2).

FIGURE 8.2 Exploring spaces during outdoor play

(Source: Amy MacDonald)

Basile (1999) highlights the particular potential of outdoor play in natural environments. As Basile explains:

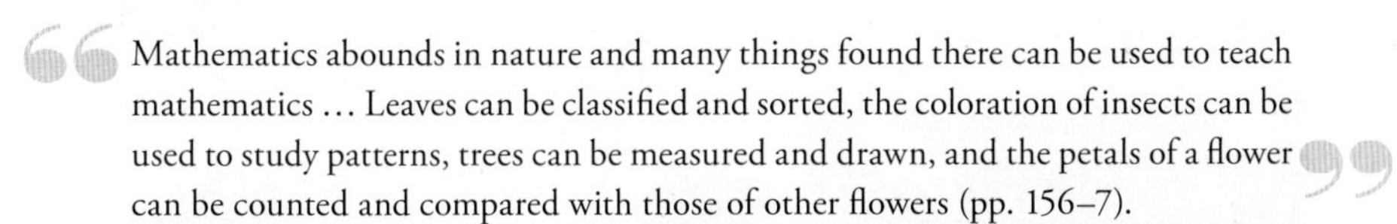

> Mathematics abounds in nature and many things found there can be used to teach mathematics … Leaves can be classified and sorted, the coloration of insects can be used to study patterns, trees can be measured and drawn, and the petals of a flower can be counted and compared with those of other flowers (pp. 156–7).

Opportunities to play with mathematical ideas in natural settings are important for children, particularly in light of the increase in children's digital and screen-based play. A balanced approach is required to help children see, and appreciate, the mathematics in a range of play situations.

Digital play

Increasingly (and perhaps controversially), young children's play involves digital technology (Arthur et al., 2015). **Digital technology** can be defined as electronic tools, systems, devices, and resources. Used in moderation, digital technologies can provide an engaging play context for exploring mathematical ideas. Australian researcher Kate Highfield has conducted extensive work around children's engagement with **digital play** (play that incorporates digital technology) and its implications for mathematical learning, specifically (Highfield, 2010). In particular, Highfield has highlighted the affordances of

Digital technology can be defined as electronic tools, systems, devices, and resources.

Digital play is play that incorporates digital technology.

simple programmable toys such as Bee-Bots (see Figure 8.3) for children's mathematical learning and problem solving. Bee-Bots are simple robots that are programmable and designed to be used on the floor (Howell & McMaster, 2022). They can be programmed either manually, or via an accompanying software program. Bee-Bots have lots of potential for enhancing mathematical activities as they engage children's spatial awareness, requiring children to consider things like position and direction when programming the Bee-Bot. For example, children might use the Bee-Bot to trace out number patterns on numbered floor mats, build obstacle courses for their Bee-Bot, or use the Bee-Bot to act out a story (Howell & McMaster, 2022). As Goodwin and Highfield (2013) explain, 'these programmable toys offer tangible interactions and provide opportunity for young learners to engage in a range of mathematical concepts and processes as they input, execute and reflect upon programs' (p. 208).

TECHNO-TOYS

Highfield (2010) has also identified how the nature of children's toys is impacted by technological developments. In particular, the development of increasingly small, accessible and cheaply manufactured digital technologies has given rise to a breed of toys that Highfield (2010) refers to as 'techno-toys'. The term **techno-toys** denotes a new generation of toys that incorporate technologies—such as embedded electronics, response systems, and microchips—in their design (Highfield, 2010). Highfield has developed a useful classification system that categorises techno-toys according to their technical features. This classification system is very helpful in identifying the range of toys that might be considered techno-toys. Highfield's classification system, and explanation of the three main categories, can be seen in Figure 8.4 and Table 8.1, respectively.

Techno-toys
are a new generation of toys that incorporate technologies—such as embedded electronics, response systems, and microchips—in their design.

FIGURE 8.3 Bee-Bot

FIGURE 8.4 Classification of techno-toys

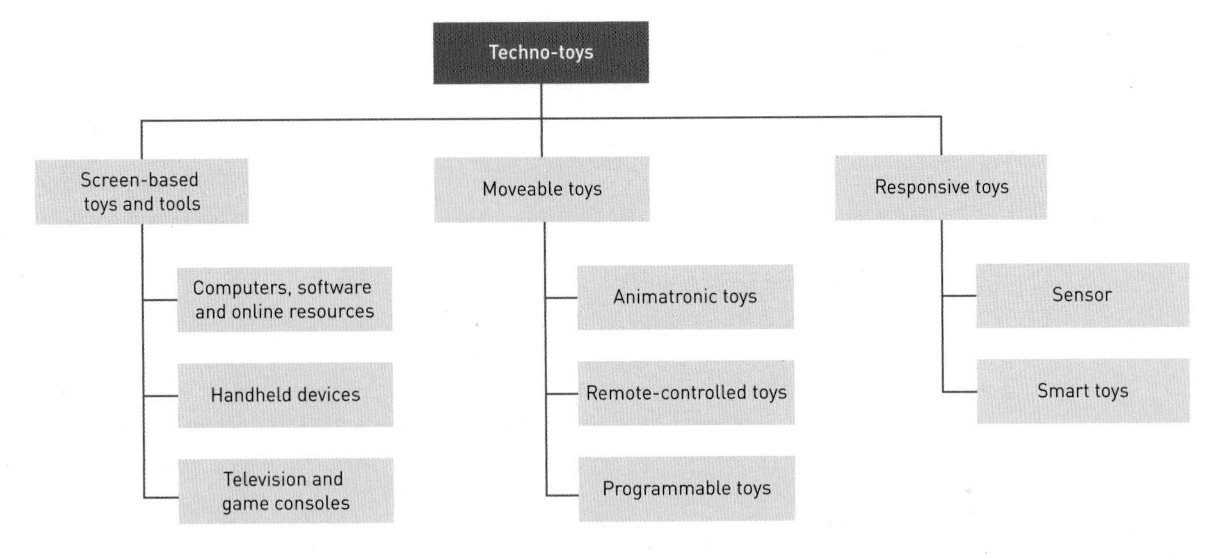

(Source: Highfield, 2010, p. 182)

TABLE 8.1 Explanation of the three categories of techno-toys

SCREEN-BASED TOYS AND TOOLS	The screen-based toys and tools category encapsulates all toys and tools that contain a screen. This includes television-based games consoles, computers and small, handheld devices, such as the Nintendo DS, Pixel Chicks and Tamagotchi. Screen-based toys and tools also include computer programs and online resources, such as Kidpix.
MOVEABLE TOYS	Moveable toys incorporate toys that travel distances and toys that are capable of 'stationary movement' (i.e. toys that don't travel, but have parts that move). Remote-control cars and programmable robotic toys are two subsets of this group; both are generally able to move two-dimensional and three-dimensional planes. Toys where only one part of the toy is capable of movement are animatronic toys. An example of this is Teddy Ruxpin; his head and facial features move, but his body remains stationary.
RESPONSIVE TOYS	Responsive toys can respond to a child's input. These toys do not have a screen and are not capable of movement. An example of this group is the Tag reader system, which is a reading pen that, when used with compatible books, responds by reading aloud. Other toys in this category respond to the child's actions, for example, the Interactive Pooh Bear, which speaks when a child squeezes its paw. Smart tools are a subset of this group. Smart toys have the capacity to adapt to the interactivity of the player; these toys, such as the Furby, appear to 'learn' as the player interacts with them.

(Source: Highfield, 2010, p. 183)

What techno-toys can you identify in your learning environment? What digital features can you identify in each of these toys—for example, does it have a screen? does it move? does it respond to you in some way? How might these techno-toys be utilised to enhance children's mathematical understandings?

Pause and reflect

APPS AND GAMES

Children's lives are rich in digital technologies, and the accessibility of mobile technologies such as smart phones and tablets means that most young children have some experience playing with apps and games on these devices. While there are certainly a number of issues associated with this type of digital play, exploration of apps and games does present a range of opportunities for developing mathematical understandings. For example, many games require children to keep track of quantities (for example, lives, points), and navigate maps. They may also be required to remember patterns or codes and make decisions based on estimation and probability. Goodwin and Highfield (2012) conducted an analysis of the 'Top Ten' paid apps located in the 'Education' section of the iTunes App store at six, six-monthly intervals, and found that many apps embed mathematical processes such as scoring and problem solving in game play (Goodwin & Highfield, 2013). However, only 15% of the Apps reviewed focused specifically on mathematical content (Goodwin & Highfield, 2013).

CURRICULUM CONNECTIONS

Early Years Learning Framework

Outcome 4: children are confident and involved learners

Children resource their own learning through connecting with people, place, technologies and natural and processed materials.

This is evident when children:

- Experiment with different technologies
- Use information and communication technologies (ICT) to investigate and problem solve.

Educators promote this learning when they:

- Introduce appropriate tools, technologies and media and provide the skills, knowledge and techniques to enhance children's learning
- Develop their own confidence with technologies available to children in the setting.

(DET, 2019, p. 40)

Music and movement

Studies show that music and movement can facilitate young children's development of mathematical abilities (Kim, 1999). The multisensory nature of music makes it a natural and inspiring way to explore mathematics with young children (Kim, 1999). Number and counting songs, in particular, are embedded in most children's everyday play activities. Singing songs and engaging in body-movement activities such as clapping hands and moving in rhythmic motions can help children explore a range of mathematical concepts and processes including counting, patterns and relationships, position and direction, sequence, repetition, and many others.

As Greenes (1999) explains, movement games require children to interpret visual stimuli and coordinate their body movements to match the movement sequence, for example, *hop, hop, jump, hop, hop, jump*. Children can also use music to explore patterns in pitch and instrumentation. For example, children can identify patterns in the pitch of individual notes (*high, high, low, low, high, high, low, low*), or create patterns with different instruments or vocal sounds (see Figure 8.5). Playing with sound can help children to imitate, complete, extend, and describe patterns, and create their own patterns (Greenes, 1999).

FIGURE 8.5 Playing with sound

(Source: Amy MacDonald)

What mathematics is used in the songs and body-movement activities you enjoy with children? What questions might you ask to assist children to notice and explore the mathematics in these songs and activities?

CURRICULUM CONNECTIONS

Early Years Learning Framework

Outcome 5: Children are effective communicators

Children engage with a range of texts and gain meaning from these texts.

This is evident when children:

- Listen and respond to sounds and patterns in speech, stories and rhymes in context
- Sing and chant rhymes, jingles and songs.

Educators promote this learning when they:

- Sing and chant rhymes, jingles and songs
- Engage children in play with words and sounds.

(DET, 2019, p. 44)

Children's drawings

Drawing presents an opportunity for children to construct and represent mathematical ideas in meaningful ways. Much can be learnt about children's understandings of size, shape, quantity, and proportion (to name a few concepts) through viewing children's own spontaneous drawings. Additionally, deeper insights can be gained by encouraging children to talk about their drawings. Placing an emphasis on listening to children while they draw is important, as the children's narratives and interpretations of their drawings can give a better picture than adults' interpretations of the drawings (Clark, 2005; Einarsdóttir, 2007).

Children can also be prompted to draw mathematical concepts as a way of revealing their experiences with, and understandings of, such concepts. For example, the invitation to 'draw a clock' is an open-ended drawing task I have used many times with very young children both as part of my teaching and as part of my research (see, for example, MacDonald, 2013; Smith & MacDonald, 2009). Children are given a blank piece of paper and are simply asked to 'draw a clock'—no further direction is given. Many children also include an accompanying narrative that reveals their different experiences with clocks in a range of contexts and gives insight into how their understandings have developed. For instance, Figure 8.6, drawn by Zofi (age five), had the following accompanying narrative: 'I did a town clock. It's a brown one. I drew numbers. It's 12 and 6 – 6 o'clock.'

Researchers have investigated the value of drawing in children's mathematics education. Lehrer, Jacobson, Kemeny, and Strom (1999) acknowledge that the graphical nature of drawing makes it a particularly useful means of accessing mathematical knowledge: 'Mathematical inscriptions seem to flow easily from children's drawings and other efforts to render the world visible' (p. 70). Woleck (2001) used children's drawings as a means of investigating children's strategies for solving open-ended mathematical problems, and found that the drawings were capturing the process by which children created meaning and understanding about mathematical concepts. In a similar vein, Carruthers and Worthington (2006) have conducted extensive studies into children's drawings, and the way in which they can use their own marks to make their own meanings. They argue that 'this allows children to more readily translate between their informal "home mathematics" and the abstract symbolism of "school mathematics"' (p. 2).

FIGURE 8.6 Zofi's drawing of a clock

(Source: Amy MacDonald)

CURRICULUM CONNECTIONS

Early Years Learning Framework

Outcome 5: Children are effective communicators

Children express ideas and make meaning using a range of media.

This is evident when children:

- Use the creative arts such as drawing, painting, sculpture, drama, dance, movement, music and storytelling to express ideas and make meaning
- Experiment with ways of expressing ideas and meaning using a range of media.

Educators promote this learning when they:

- Build on children's family and community experiences with creative and expressive arts
- Provide a range of resources that enable children to express meaning using visual arts, dance, drama and music.

(DET, 2019, p. 45)

Storybooks

Children's storybooks provide opportunities for noticing, exploring, and talking about mathematical ideas. There are many children's books that are written specifically to develop mathematical ideas; however, there is mathematics in most children's storybooks (Perry & Gervasoni, 2012). Some storybooks contain 'overt' mathematical elements such as counting sequences or numerals, while others contain more subtle mathematical elements. For example, stories with repeated narrative elements, rhythms, or rhymes help to develop concepts of pattern and structure.

Stories also provide opportunities to compare and classify, explore relationships, and identify objects and shapes. Storybooks can make mathematical concepts relevant to children because stories provide children with problem situations and solutions in a narrative context (Hong, 1999). Additionally, storybooks provide a context that is interesting and meaningful to children (Hong, 1999).

Dunphy (2020) offers a useful framework for selecting and using storybooks for early mathematics education (Table 8.2).

TABLE 8.2 A framework for selecting and using storybooks in early childhood mathematics education

PRESENTATION	CONTENT	CHILDREN'S PARTICIPATION
1. How is the content presented? Explicit/implicit (i.e. not directly explained) 2. Is there mathematical content visible but not related to the story itself? 3. Will the content engage the children in a mathematically meaningful way? 4. How does the storybook relate mathematics to children's lives, interests, and experiences? 5. How might the mathematical content be integrated with other areas of learning, for example, through pretend play? 6. How might the storybook make understanding possible at different levels; offer multiple layers of meaning; anticipate future concept development?	1. What mathematical processes does the book support? For example: – using mathematical language – reflecting on mathematical activities – solving mathematical problems – mathematical reasoning. 2. What key ideas are encountered? – Sets – Number sense – Counting – Number operations – Pattern – Measurement – Data analysis – Spatial relationships – Shape. 3. What strand of mathematics can be developed? For example: – Number – Algebra – Shape and space – Measurement – Data 4. What aspect of the content is addressed?	1. What activities will arise from the storybook? 2. What communication modes might children use to engage mathematically with the activities? 3. How might children demonstrate receptive understanding of the mathematics? 4. What mathematical language will be developed as a result of engagement with the story? 5. What questions will be key to provoking children's participation and discussion? 6. What kinds of discussion might arise, and how will the children be supported to mathematise? 7. How will children's engagement in mathematical discussion be maximised?

(Source: Dunphy, 2020, adapted from van den Heuvel-Panhuizen & Elia, 2012)

Opinion Piece 8.1

DR LORRAINE GAUNT AND DR MELLIE GREEN

Using high quality picture books, such as those shortlisted each year by the Children's Book Council of Australia (CBCA), can harness children's enjoyment of reading to support their engagement in mathematical learning. Engagement in children's literature can support the development of students' interest in mathematics throughout schooling to develop the numerate citizens we all wish to see. While there are many children's books written for the purpose of teaching mathematical concepts such as *A dozen dizzy dinosaurs* by James Burnett and Calvin Irons, and other well-known books frequently used to support some mathematical understandings, such as *The very hungry caterpillar* by Eric Carle and *Counting on Frank* by Rod Clements, we believe that most children's picture books can be used to promote mathematical thinking. You just need to read the books with a mathematical lens. For example, *Stellarphant* by James Foley has many links with concepts of measurement, geometry and space. By using a book shortlisted for the CBCA Book of the Year awards, teachers can be assured that the literature they are using is likely to promote aesthetic enjoyment in their young students as well as support both their literacy and numeracy development.

Lorraine Gaunt is a lecturer in mathematics education and inclusive education at Charles Sturt University, NSW. She has worked extensively in Queensland schools both teaching secondary mathematics and as Head of Special Education Services supporting the inclusion of students with additional needs into mainstream classrooms. Lorraine is currently researching in the area of enabling teachers to use quality children's literature to support student engagement and understanding of mathematics from a Reading for Enjoyment pedagogy. Lorraine is particularly interested in including all students in the school community, particularly in the mathematics classroom. In 2022, Lorraine was awarded the Early Career Researcher award at the Mathematics Education Research Group of Australasia's annual conference.

Mellie Green is a lecturer at Southern Cross University (Gold Coast Campus) and has been a primary school teacher for over 25 years, with 20 of those in the classroom and teacher-librarianship, and five in curriculum leadership. She completed her PhD in 2021. Her doctoral research explored student engagement in reading for enjoyment in the upper primary years. Her areas of research passion are use of children's literature in the primary classroom, reading instruction, curriculum and pedagogy. She is a Literacy and English teaching lecturer with a love of high-quality children's books. Mellie is also an active member of the Departing Radically in Academic Writing (DRAW) group.

Learn more
Go online to hear Dr Lorraine Gaunt and Dr Mellie Green read this Opinion Piece.

CURRICULUM CONNECTIONS

Early Years Learning Framework

Outcome 5: Children are effective communicators

Children engage with a range of texts and gain meaning from these texts.

This is evident when children:

- Listen and respond to sounds and patterns in speech, stories and rhymes in context
- View and listen to printed, visual and multimedia texts and respond with relevant gestures, actions, comments and/or questions.

Educators promote this learning when they:

- Read and share a range of books and other texts with children
- Talk explicitly about concepts such as rhyme and letters and sounds when sharing texts with children
- Incorporate familiar family and community texts and tell stories.

(DET, 2019, p. 44)

The Raising Children Network (2017) recommends the following children's storybooks for exploring mathematical ideas:

- *The Very Hungry Caterpillar*, by Eric Carle
- *One Fish, Two Fish, Red Fish, Blue Fish*, by Dr Seuss
- *Ten Little Ladybugs*, by Melanie Gerth
- *Counting Kisses*, by Karen Katz
- *Ten Little Fingers and Ten Little Toes*, by Mem Fox
- *One Woolly Wombat*, by Kerry Argent
- *Ten Little Dinosaurs*, by Mike Brownlow

Of course, there are many others that might be added to this list. A personal favourite is *Where is the Green Sheep?* by Mem Fox and Judy Horacek (2004). The delightfully predictable narrative structure assists children to recognise rhythmic patterns, rhyme, and repeating elements within the text. Furthermore, the story explores spatial concepts such as position and direction, measurement concepts associated with size, mass, and comparison, as well as notions of similarity/difference and opposites.

What other storybooks can you think of that might be used for exploring mathematics with young children?

Chapter summary

This chapter has outlined some of the many opportunities for learning about mathematics that exist in children's everyday lives. Play contexts provide meaningful ways of exploring mathematics, as children can use mathematics in their own ways and for their own purposes. The activities canvassed in this chapter can easily be embedded in more early childhood mathematics education programs as a way of fostering links between children's informal uses of mathematics and more formal applications of mathematical concepts and processes.

The specific topics we covered in this chapter were:

- mathematics in play
- outdoor play
- digital play
- techno-toys
- apps and games
- music and movement
- children's drawings
- storybooks.

FOR FURTHER REFLECTION

What is your favourite children's story? Re-read the story through a mathematical lens and think about the following:

1 What mathematical ideas do you notice in the story?
2 What strategies might you use to help a child notice these mathematical ideas?
3 How could the story be used as the basis for a mathematical learning experience in the classroom?
4 In what ways might the mathematical concepts embedded in the story be extended?

FURTHER READING

Basile, C.G. (1999). The outdoors as a context for mathematics in the early years. In J.V. Copley (ed.), *Mathematics in the early years* (pp. 156–61). Reston, VA: National Council of Teachers of Mathematics.

Carruthers, E., & Worthington, M. (2006). *Children's mathematics: Making marks, making meaning* (2nd ed.). London: SAGE.

Highfield, K. (2010). Possibilities and pitfalls of techno-toys and digital play in mathematics learning. In M. Ebbeck & M. Waniganayake (eds), *Play in early childhood education: Learning in diverse contexts* (pp. 177–96). South Melbourne, VIC: Oxford University Press.

Hong, H. (1999). Using storybooks to help young children make sense of mathematics. In J.V. Copley (ed.), *Mathematics in the early years* (pp. 162–8). Reston, VA: National Council of Teachers of Mathematics.

Kim, S.L. (1999). Teaching mathematics through musical activities. In J.V. Copley (ed.), *Mathematics in the early years* (pp. 146–50). Reston, VA: National Council of Teachers of Mathematics.

Smith, T., & MacDonald, A. (2009). Time for talk: The drawing-telling process. *Australian Primary Mathematics Classroom, 14*(3), 21–6.

09 Working Mathematically through Powerful Processes

Learn more
Go online to see a video from Paige Lee explaining the key messages of this chapter.

CHAPTER OVERVIEW

This chapter will explore the powerful processes that children use to develop and apply mathematical knowledge. Engagement with powerful processes allows children to work mathematically and develop dispositions for mathematical learning. This chapter will also explore how processes of questioning and problem solving provide opportunities for mathematical learning and development.

In this chapter, you will learn about the following topics:

- working mathematically
- Bishop's Mathematical Activities
  - counting
  - measuring
  - locating
  - designing
  - playing
  - explaining
- identifying Bishop's Mathematical Activities in children's play
- dispositions for mathematical learning
- questioning
- using questions to guide mathematical thinking
- problem solving.

KEY TERMS

- Bishop's Mathematical Activities
- Concepts
- Counting
- Designing
- Dispositions
- Explaining
- Fluency
- Locating
- Measuring
- Playing
- Problem
- Problem solving
- Processes
- Reasoning
- Understanding
- Working mathematically

Introduction

Children acquire understanding of mathematics concepts through active involvement with their environments (Lind, 1998). As such, the notions of *action* or *activity* become essential for conceptual development. It is therefore important to consider mathematics in relation to both *concepts* and *processes*. Concepts and processes are separate, but also interrelated, ideas. **Concepts** represent the 'content', the mathematical knowledge being developed—the *what*. Concepts can be thought of as the building blocks of knowledge, and they allow people to organise and categorise information (Stelzer, 2005). **Processes** are the ways in which we explore concepts; they are the actions, the verbs ('doing words')—the *how*. For example, one-to-one correspondence is a *concept* developed through the *process* of counting. Early years educators can actively promote children's engagement with such processes to assist children's conceptual development (MacDonald, 2015a).

Concepts
are the building blocks of knowledge that allow people to organise and categorise information.

Processes
are the actions through which concepts are explored.

Using the processes of mathematics is often described as **working mathematically**. This term is used in the *Australian Curriculum: Mathematics* (ACARA, 2022) to encapsulate the proficiencies of *understanding*, *fluency*, *problem solving*, and *reasoning*, which describe how mathematical content is explored or developed.

Working mathematically
means using the processes of mathematics.

Powerful processes operate as a mechanism that provides the actions and activities to support children's investigative processes in mathematics education (MacDonald, 2015a). In mathematics education, we often begin with a problem to be solved, followed by a systematic process of inquiry where the problem is explored and further investigated (Aitken, Hunt, Roy, & Sajfar, 2012). These problem-solving processes allow children to engage with mathematical questioning and use a range of strategies to answer their questions. Indeed, questioning plays an important role in supporting the development of powerful processes. It could be argued that the key to children's learning lies in the questions the children ask and the questions that are asked of them, because asking questions helps to foster curious and creative dispositions for learning in mathematics (MacDonald, 2015a).

This chapter provides an overview of the powerful processes that can be developed in the early childhood years and presents examples of how these powerful processes can foster children's exploration of mathematical concepts.

Opinion Piece 9.1

PAIGE LEE

Coming originally from a history education background, and having made my way into early childhood mathematics, understanding how children learn is something I value highly. Through knowing and understanding the processes by which children engage with and learn about different concepts, educators can open up an array of transferable skills, knowledge, and understandings that are essential for mathematical learning now and in the future, and that contribute to a range of content areas and life experiences. Fostering processes in our children gives them the tools to understand

Learn more
Go online to hear Paige Lee read this Opinion Piece.

their world—content knowledge is more readily accessible, but the skills needed to be critical, active, and independent learners are attributes that I feel we as educators can and should support wholeheartedly.

Paige Lee is Research Associate, Sessional Academic, and Quality Assurance Officer at Charles Sturt University, Albury–Wodonga, Australia. She is also a K–12 teacher, specialising in history education. She has worked on a number of early childhood projects, including transition to school and early years mathematics projects. Paige currently teaches in the early childhood mathematics education subjects in Charles Sturt University's Bachelor of Education (Birth to Five Years) degree program.

CURRICULUM CONNECTIONS

Early Years Learning Framework

Outcome 4: Children are confident and involved learners

Children develop a range of skills and processes such as problem solving, inquiry, experimentation, hypothesising, researching and investigating.

This is evident when children:

- Apply a wide variety of thinking strategies to engage with situations and solve problems, and adapt these strategies to new situations
- Make predictions and generalisations about their daily activities, aspects of the natural world and environments, using patterns they generate or identify and communicate these using mathematical language and symbols
- Use reflective thinking to consider why things happen and what can be learnt from these experiences.

Educators promote this learning when they:

- Plan learning environments with appropriate levels of challenge where children are encouraged to explore, experiment and take appropriate risks in their learning
- Provide babies and toddlers with resources that offer challenge, intrigue and surprise, support their investigations and share their enjoyment
- Provide experiences that encourage children to investigate and solve problems
- Encourage children to use language to describe and explain their ideas
- Provide opportunities for involvement in experiences that support the investigation of ideas, complex concepts and thinking, reasoning and hypothesising
- Join in children's play and model reasoning, predicting and reflecting processes and language
- Listen carefully to children's attempts to hypothesise and expand on their thinking through conversation and questioning.

(DET, 2019, p. 38)

Working mathematically

As identified in the Introduction to this chapter, the *Australian Curriculum: Mathematics* (ACARA, 2022) encourages children to work mathematically by using four proficiencies:

1 understanding
2 fluency
3 problem solving
4 reasoning.

As ACARA explain, these proficiencies describe the actions in which children can engage when learning and using mathematical content. Furthermore, they indicate the breadth of mathematical actions that teachers can emphasise. The four proficiencies, as described by ACARA, are presented in Table 9.1.

TABLE 9.1 Proficiencies for working mathematically

UNDERSTANDING	Students build a robust knowledge of adaptable and transferable mathematical concepts. They make connections between related concepts and progressively apply the familiar to develop new ideas. They develop an understanding of the relationship between the 'why' and the 'how' of mathematics. Students build understanding when they connect related ideas, when they represent concepts in different ways, when they identify commonalities and differences between aspects of content, when they describe their thinking mathematically and when they interpret mathematical information.
FLUENCY	Students develop skills in choosing appropriate procedures; carrying out procedures flexibly, accurately, efficiently and appropriately; and recalling factual knowledge and concepts readily. Students are fluent when they calculate answers efficiently, when they recognise robust ways of answering questions, when they choose appropriate methods and approximations, when they recall definitions and regularly use facts, and when they can manipulate expressions and equations to find solutions.
PROBLEM SOLVING	Students develop the ability to make choices, interpret, formulate, model and investigate problem situations, and communicate solutions effectively. Students formulate and solve problems when they use mathematics to represent unfamiliar or meaningful situations, when they design investigations and plan their approaches, when they apply their existing strategies to seek solutions, and when they verify that their answers are reasonable.
REASONING	Students develop an increasingly sophisticated capacity for logical thought and actions, such as analysing, proving, evaluating, explaining, inferring, justifying and generalising. Students are reasoning mathematically when they explain their thinking, when they deduce and justify strategies used and conclusions reached, when they adapt the known to the unknown, when they transfer learning from one context to another, when they prove that something is true or false, and when they compare and contrast related ideas and explain their choices.

(Source: ACARA, 2022)

Understanding means having a robust knowledge of adaptable and transferable mathematical concepts.

Fluency means carrying out procedures flexibly, accurately, efficiently, and appropriately.

Problem solving refers to formulating, investigating, and resolving problems, and communicating solutions.

Reasoning refers to capacity for logical thought and actions.

Bishop's Mathematical Activities

Bishop's Mathematical Activities are the processes of counting, measuring, locating, designing, playing, and explaining.

There are many mathematical process models proposed in the scholarly literature. One such model that is highly appropriate to the early childhood years is **Bishop's Mathematical Activities** (Bishop, 1988). As outlined in Chapter 5, Alan Bishop was a mathematician who observed the mathematics of cultures around the world. Based on his observations, Bishop (1988) proposed a model of 'Six Universal Mathematical Activities', which represents the mathematical processes common to learning mathematics, regardless of context. Bishop's six processes are as follows:

1 counting
2 measuring
3 locating
4 designing
5 playing
6 explaining.

Bishop's six activities are described below and are summarised in Table 9.2.

TABLE 9.2 Bishop's Mathematical Activities

COUNTING	The process of expressing numerical quantifiers and qualifiers, which is stimulated by, and in turn affects, classifying and pattern-seeking
MEASURING	The process of ascertaining or calculating quantities or entities that cannot be counted or located spatially, and also involves comparing, ordering, and quantifying
LOCATING	The process of positioning oneself and other objects in the spatial environment, which includes expressions of position, shape, boundedness, continuousness, direction, and physical or temporal space (i.e. time)
DESIGNING	The process of expressing a symbolic plan, structure, or shape (or surface of a shape), which bridges the relationship between an object (real or imagined) and its purpose, and does not need to result in a finished product—that is, the design does not actually have to be made
PLAYING	The process that imitates or imaginatively recreates social, concrete, or abstract models of reality
EXPLAINING	The process of expressing factual or logical aspects of ideas, questions, experiences, events, or relationships between phenomena

(Adapted from Bishop, 1988)

COUNTING

Counting is the process of expressing numerical quantifiers and qualifiers.

Counting is the process of expressing numerical quantifiers and qualifiers. It is an activity that is stimulated by, and in turn affects, classifying and pattern-seeking (Macmillan, 1995). Examples of the ways in which young children might demonstrate the process of counting include: 'I've got lots of cars and trucks'; and 'Two more teddy bears for your picnic' (MacDonald, 2015a). In baby and toddler settings, counting expressions are evident in songs and nursery rhymes, and in everyday transitions and routines such as meal times.

Measuring is the process of ascertaining or calculating quantities or entities that cannot be counted or located spatially.

MEASURING

Measuring is the process of measuring quantities or entities that cannot be counted or located spatially, and also involves comparing, ordering, and quantifying (Macmillan, 1995). Examples of the ways in which young children might demonstrate the process of measuring include: 'This is heavy'; and 'Watch me jump as high as I can!' (MacDonald, 2015a). Babies and toddlers might engage with measurable attributes such as time when they recognise routines and certain times of the day, or demonstrate awareness of quantity when indicating they want 'more' of something (for example, 'more milk').

LOCATING

Locating involves positioning oneself and other objects in the spatial environment (Macmillan, 1995). It includes appreciation of position, shape, boundedness, continuousness, direction, and physical or temporal space (i.e. time). Examples of the ways in which young children might demonstrate the process of locating include: 'This is the way up and this is the way down'; and 'It needs to cook in the oven for a long time' (MacDonald, 2015a). Babies and toddlers explore location by positioning themselves in space, for example, crawling under, over, through; and recognising different times of the day.

Locating
is the process of positioning oneself and other objects in space.

DESIGNING

Designing is an expression of a symbolic plan, structure, or shape (or surface of a shape). This process bridges the relationship between an object (real or imagined) and its designated purpose, and does not need necessarily to result in the production of a finished product—that is, the design does not actually have to be made (Macmillan, 1995). Examples of the ways in which young children might demonstrate the process of designing include: 'Look, I'm making a tunnel!'; and 'We have to make a house for all the teddies' (MacDonald, 2015a). Babies and toddlers might build with blocks, arrange pillows, or move food around on a plate or table to express an intended design, shape, or structure.

Designing
is the process of expressing a symbolic plan, structure, or shape.

PLAYING

Playing is a process that imitates or imaginatively recreates social, concrete, or abstract models of reality (Macmillan, 2009). Bishop (1988) considered play to be a crucial activity for mathematical development. Examples of the ways in which young children might demonstrate the process of playing include: 'All the teddies are having a rest because it is a hot day'; and 'I am the fairy queen and I have a big castle!' (MacDonald, 2015a). Babies and toddlers might mimic noises or movements in their play as a way of imitating or recreating their observations of reality.

Playing
is the process of imitating or recreating social, concrete, or abstract models of reality.

EXPLAINING

Explaining is the process of expressing factual or logical aspects of ideas, questions, experiences, events, or relationships between phenomena (Macmillan, 2009). Examples of the ways in which young children might demonstrate the process of explaining include: 'You put them in the fridge and when they're cold you eat them'; and 'If we put these two together it will be bigger' (MacDonald, 2015a). Babies and toddlers engage in explaining when they explore and respond to logical aspects of ideas, questions, events, or relationships.

Explaining
is the process of expressing factual or logical information.

IDENTIFYING BISHOP'S MATHEMATICAL ACTIVITIES IN CHILDREN'S PLAY

To assist you in recognising Bishop's processes in children's activity, Table 9.3 presents an observation of an interaction between a child and an educator and analyses the observation in relation to Bishop's six Mathematical Activities.

TABLE 9.3 Identifying Bishop's Mathematical Activities in a child's play
Context: Two-year-old Ava (A) is engaged in solitary play with Duplo blocks. Her educator (E) comes over to inquire as to what Ava is constructing.

OBSERVATION	MATHEMATICAL ACTIVITIES
[A has a large container of Duplo blocks, of varying colours and sizes. She is standing next to a low table and is arranging the blocks on the table. One by one, she begins connecting the blocks together by placing them on top of one another.]	*Locating*: arranging the blocks on the table
E: What are you making, Ava?	
A: Big tower.	*Playing*: creating a model of reality *Designing*: expressing the intended design of the tower *Measuring*: expressing the height of the tower *Explaining*: responding to E's question
[A hands a block to E.]	
A: This one.	*Designing*: directing E to contribute to the design of the tower
E: Would you like me to put this block somewhere?	
[A points to the top of his structure.]	*Designing*: directing E to contribute to the design of the tower *Locating*: indicating the spatial location for E's block
A: Put there.	*Locating*: expressing where E's block needs to be positioned
[E places the block in the position indicated by A. A pauses and looks at the block for a moment.]	
A: Not this one. [A selects another, smaller, block and hands it to E.] This one.	*Measuring*: comparing the sizes of the blocks *Designing*: refining the design of the structure
E: Would you like me to put this block there instead?	
A: Put there.	*Locating*: expressing where E's block needs to be positioned
[E removes the previous block and puts the new block in its place.]	
A: That's better.	*Designing*: confirming that the placement of E's block fits with the design

Dispositions for mathematics learning

Dispositions are characteristics that encourage children to respond in particular ways to learning opportunities (Aitken et al., 2012). Outcome 4 of the *Early Years Learning Framework for Australia* (EYLF; DET, 2019) identifies nine dispositions for learning that are just as important in mathematics education as they are in other areas of early childhood learning, wellbeing, and development. Table 9.4 provides an overview of the nine dispositions and how these relate to early childhood mathematics education.

TABLE 9.4 Dispositions for working mathematically

CURIOSITY	Children are naturally curious, and as adults we need to notice children's curiosity but know when to stay quiet and when to speak. When appropriate, we can model suitable mathematical vocabulary to further stimulate children's curiosity.
COOPERATION	Many mathematical investigations require children to work together with others (this may be other children or an educator) towards an outcome. Cooperation means working with others, listening to their ideas, rejoicing in the group's achievements, sharing work, and taking turns.
CONFIDENCE	It is important to build children's confidence in mathematics by providing encouragement for their involvement in learning experiences, being available to support their contributions, encouraging further involvement, and building on their interest and findings.
ENTHUSIASM	When we show interest in mathematics, and provide opportunities for children to investigate, it is likely that children's enthusiasm will follow.
CREATIVITY	Children are naturally creative, and this creativity can be expressed in mathematics education by encouraging children to pursue their own ideas and work out their own answers by investigating different possibilities. Children can come up with exciting and unique answers to their mathematical questions.
COMMITMENT	Educators will build children's commitment to learning about mathematics if they allow extended time for children to explore an idea without constant changes. The provision of rich and stimulating learning opportunities will entice children to commit their time and energy to the investigation.
PERSISTENCE	Children can develop persistence through inquiry-based and investigative mathematical learning experiences. Trialling different solutions to problems can encourage children to persist in order to achieve a positive outcome.
IMAGINATION	Young children have wonderful imaginations and can come up with exciting solutions to mathematical questions based on their current understandings about the world. Sometimes we need to let children hold on to their 'magical' beliefs about the world around them—and when they are ready, they will ask questions that lead to more mathematical understandings about the world.
REFLEXIVITY	Children demonstrate reflexivity when they reflect on their learning and relationships within mathematics. Through discussions about what has happened, we can help children to review and consolidate the mathematics concepts, processes, and vocabulary they have learnt.

(Adapted from Aitken et al., 2012, p. 21).

CURRICULUM CONNECTIONS

Early Years Learning Framework

Outcome 4: Children are confident and involved learners

Children develop dispositions for learning such as curiosity, cooperation, confidence, creativity, commitment, enthusiasm, persistence, imagination and reflexivity.

This is evident when children:

- Are curious and enthusiastic participants in their learning
- Use play to investigate, imagine and explore ideas
- Participate in a variety of rich and meaningful inquiry-based experiences.

Educators promote this learning when they:

- Respond to children's displays of learning dispositions by commenting on them and providing encouragement and additional ideas
- Encourage children to engage in both individual and collaborative explorative learning processes
- Model inquiry processes, including wonder, curiosity and imagination, try new ideas and take on challenges.

(DET, 2019, p. 37)

Questioning

Questioning is critical in the development of powerful processes for mathematics learning. Questioning helps to prompt actions that allow children to engage with conceptual ideas. However, in order for such engagement to occur, the importance of asking 'good' questions must not be underestimated. A 'good' question can be considered as having three features:

1. It requires more than recall or reproduction of a skill.
2. It has an educative component—that is, the child will learn from attempting to answer the question, and the educator will learn about the child from their attempt.
3. It is to some extent open—there may be several acceptable answers.

To assist you in formulating good questions, and also to help you guide children's own questioning, it is important to understand the specific characteristics of good questions. Good questions require children to:

- manipulate prior information
- state an idea in their own words
- find a solution to a problem
- observe and describe an event or an object
- compare two or more objects
- give examples
- explain their thinking
- apply ideas to new situations

- compare and find relationships
- make predictions or inferences
- make a judgment.

In short, good questions *challenge children to think*. The deeper the question posed, the more the child will need to think, and the greater the potential for conceptual development. However, when asking cognitively challenging questions, it is important to ask—then wait. The longer you wait, the more the child's thinking can take place.

When posing challenging questions for children, it is also important to handle their answers with sensitivity. Ensure that you acknowledge all responses, and don't reject an answer if it is not what you expected. Inappropriate responses can be addressed by clarifying your question or rephrasing the question. You might also consider extending the question further, or asking others for their responses, to help challenge the child's thinking.

USING QUESTIONS TO GUIDE MATHEMATICAL THINKING

Beyond setting up experiences and equipment, educators must plan for how they can use questions and prompts to guide and extend children's mathematical thinking during play and investigation. These questions might be used in a spontaneous manner during a child-initiated experience, but you might also use questions as the basis for your planning of educator-instigated experiences. As Harlan and Rivkin (2012, pp. 34–5) explain, open-ended questioning can serve many purposes, including:

- *Instigating discovery*—an activity becomes a discovery challenge when it is initiated as a question to answer.
- *Eliciting predictions*—it can be helpful to encourage children's predictions before children investigate an idea.
- *Probing for understanding*—careful questioning helps to uncover children's conceptual understandings.
- *Promoting reasoning*—careful questioning can also help to elicit children's thought processes.
- *Promoting reasoning*—we can ask questions which encourage children to explain and justify their ideas.
- *Serving as a catalyst*—sometimes a question can be a catalyst that sparks interest in an investigation.
- *Encouraging creative thinking and reflection*—questioning can help children to think about their learning and make connections to other experiences.
- *Reflecting on feelings*—questions can be used to identify particular concepts or experiences that capture children's interest.

Can you think of two examples of mathematical questions that you might ask a three-year-old child? Now, can you think of two examples of mathematical questions that a three-year-old child might ask you?

There are a number of specific question types that can be used to great effect in mathematical learning experiences. Questions can be grouped according to the following categories:

- exemplify, specialise
- complete, delete, correct
- compare, sort, organise
- change, vary, reverse, alter
- generalise, conjecture
- explain, justify, verify, convince, refute.

Table 9.5 gives examples of questions that fit within each of the categories.

TABLE 9.5 Categories of questions and some examples

EXEMPLIFY, SPECIALISE	Can you give me other examples of ...? Can you show, choose, find, describe an example of ...? What makes ... an example? Can you find a counter example of ...? Are there any special examples of ...?
COMPLETE, DELETE, CORRECT	What are you doing? ... Okay, and ... How will we do that? I have a problem, can you help? What do we need to do next? What can we add, remove, alter? Tell me what is wrong with ...? What needs to be changed so that ...?
COMPARE, SORT, ORGANISE	How can we sort these? What is the same/different about ...? What is the best way to ...? Do these belong together?
CHANGE, VARY, REVERSE, ALTER	What if ...? If this is the answer, what was the question? Can we do this differently? What is the quickest/easiest way to do this? Can we change this so that ...?
GENERALISE, CONJECTURE	What happens in general? Is it always/sometimes/never ...? Can you describe all possible ... as simply as you can? What can change and what has to stay the same so that ... is still true?
EXPLAIN, JUSTIFY, VERIFY, CONVINCE, REFUTE	Can you explain why ...? Can you give a reason ... (for using or not using ...)? How can we be sure that...? Tell me what is wrong with...? Is it ever false or is it always true? How is ... used in ...? (Explain the role or use of ...) Convince me that ...

(Source: MacDonald, 2015a, p. 41)

How might you base your planning around one or more of these question types so as to elicit different aspects of children's mathematical thinking?

Pause and reflect

Problem solving

Learning to problem solve is an essential feature of children's mathematical development (Geary, 1994). Skinner (1990) defined a **problem** as being a question that engages someone in searching for a solution. More specifically, working with problems involves investigation, questioning, trial and error, divergent thinking, and decision making. Furthermore, working with problems involves selecting and using a range of strategies and tools to explore and solve the problem.

A **problem** is a question that engages someone in searching for a solution.

We problem solve all the time. For example, what time did you wake up this morning? Did you choose a specific time to wake up? Did you have a reason to wake up at this time? When you went to bed last night, you were faced with the problem, *'What time do I need to wake up in the morning?'*

Pause and reflect

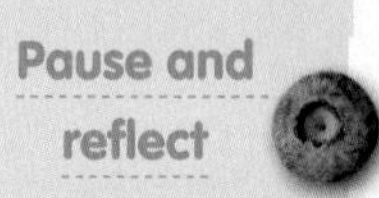

Problem solving can be considered as consisting of five stages:

1 Identify the problem.
2 Pose the problem as a question.
3 Develop a procedure for solving the problem.
4 Conduct the procedure.
5 Draw conclusions.

To illustrate each of these five stages, Table 9.6 presents a worked example of the problem, 'What time do I need to wake up in the morning?'

TABLE 9.6 Five stages of problem solving

1. Identify the problem	I need to decide what time to wake up in the morning.
2. Pose the problem as a question	What time do I need to wake up in the morning?
3. Develop a procedure for solving the problem	I will think about what I need to do in the morning. What time do I need to be at work? How long do I need to get ready? How long will it take me to get to work?
4. Conduct the procedure	I need to be at work at 8 am. It will take me 30 minutes to get ready, and 15 minutes to drive to work.
5. Draw conclusions	I need to wake up at 7.15 am.

Problem posing and problem solving begins from a very young age. You can see it when a baby follows a moving object with their eyes or looks you up and down. You see it when a baby grabs their feet, or an object, and plays with them and moves them in different ways.

A crawling infant with a toy, such as in Figure 9.1, might pose the question: 'How can I bring my toy with me?' They may not be able to verbalise that they are problem solving, but we can observe it. These sorts of perceptual, visual, and kinaesthetic investigations are very important for developing children's capacity to problem pose and problem solve, and these investigations should be encouraged by adults or other children interacting with the child.

FIGURE 9.1 An infant problem posing and problem solving

(Source: Amy MacDonald)

As children's vocabulary increases, they are able to verbalise the processes of problem posing and problem solving, and use language to ask questions, make predictions, articulate their hypotheses, give reasons for their predictions and solutions, and reflect on their learning. There are a number of things that educators can do to assist children in developing their problem-solving skills, and these should be considered when planning mathematical learning experiences (MacDonald, 2015a):

- Young children learn best when they are given frequent opportunities to solve problems that are meaningful to them—those that arise in their day-to-day life.
- Provide opportunities for hands-on investigations. Offer children interesting items to explore, and also rotate your materials to keep them fresh and thought provoking.
- Foster creative and critical thinking skills by inviting children to use items in new and diverse ways.
- Encourage children's suggestions and solutions. Promote brainstorming by asking open-ended questions.
- Allow children to find their own solutions. Offer help if they become frustrated, but don't solve their problems for them.

CURRICULUM CONNECTIONS

Early Years Learning Framework

Outcome 4: Children are confident and involved learners

Children transfer and adapt what they have learnt from one context to another.

This is evident when children:

- Use the processes of play, reflection and investigation to solve problems
- Apply generalisations from one situation to another
- Try out strategies that were effective to solve problems in one situation in a new context.

Educators promote this learning when they:

- Support children to construct multiple solutions to problems and use different ways of thinking.

(DET, 2019, p. 39)

Chapter summary

This chapter has presented the powerful processes that children use to explore and develop mathematical concepts. Understanding *processes* helps educators to recognise, support, and plan for mathematical *activity*. Bishop's model of Mathematical Activities is useful for conceptualising mathematical processes and highlights the ways in which children actively construct mathematical concepts. Educators can employ a range of pedagogical actions to encourage children to use mathematical processes in meaningful and purposeful ways.

The specific topics we covered in this chapter were:

- working mathematically
- Bishop's Mathematical Activities
- counting
- measuring
- locating
- designing
- playing
- explaining
- identifying Bishop's Mathematical Activities in children's play
- dispositions for mathematics learning
- questioning
- using questions to guide mathematical thinking
- problem solving.

FOR FURTHER REFLECTION

Consider Figure 9.2.

FIGURE 9.2 Water play

(Source: Amy MacDonald)

1 What powerful processes might be occurring in the play?
2 Which of Bishop's Mathematical Activities can you identify?
3 What problems might the children be posing and solving?
4 What questions might the children be exploring through their play?
5 What questions might you, as the educator, ask of the children?

FURTHER READING

MacDonald, A. (2015a). Powerful processes. In *Investigating mathematics, science and technology in early childhood* (pp. 14–30). South Melbourne, VIC: Oxford University Press.

Macmillan, A. (2009). Shared contexts for teaching and learning numeracy. In *Numeracy in early childhood: Shared contexts for teaching and learning* (pp. 20–33). South Melbourne, VIC: Oxford University Press.

10 Transitions to School

Learn more
Go online to see a video from Paige Lee explaining the key messages of this chapter.

CHAPTER OVERVIEW

This chapter focuses on matters surrounding children's transition from prior-to-school mathematics education to school mathematics education. Transition to school is an important time in a child's life, and this chapter explores how educators can support children's mathematics learning across the transition from their prior-to-school setting to their primary school setting.

In this chapter, you will learn about the following topics:

- transition to school
- opportunities
- aspirations
- expectations
- entitlements
- principles of effective transitions practices
- pedagogies of educational transitions
- communicating across the transition to school
- transition statements.

KEY TERMS

- Aspirations
- Entitlements
- Expectations
- Opportunities
- Pedagogies of educational transitions
- Transitions statements
- Transition to school

AMY MACDONALD

Introduction

Transition to school is an important time in any young child's life, and early childhood educators have an important role to play in supporting this transition. In particular, educators must actively support, recognise, and communicate children's learning—including children's *mathematics* learning—across this transition from the prior-to-school setting to the primary school setting. Transitioning children's mathematics education brings many opportunities, but also many challenges in terms of curriculum, pedagogy, relationships, and communication. This is recognised in *Belonging, Being and Becoming: The Early Years Learning Framework for Australia* (EYLF; DET, 2019), which states:

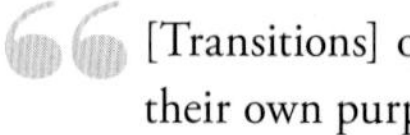

> [Transitions] offer opportunities and challenges. Different places and spaces have their own purposes, expectations and ways of doing things. Building on children's prior and current experiences helps them to feel secure, confident and connected to familiar people, places, events and understandings. Children, families and early childhood educators all contribute to successful transitions between settings (p. 19).

The focus of this chapter is perhaps best summarised by Perry, MacDonald, and Gervasoni (2015, pp. 8–9) who outline a number of key themes underpinning mathematics and transitions to school—these being:

- recognition of the mathematical, and other, strengths that all participants in the transition to school bring to this period of a child's life
- recognition of the opportunities provided by transition to school for young children's mathematics learning
- the importance of partnerships among adults, and among adults and children, for effective school transitions and mathematics learning and teaching
- the critical impact of the expectations of all involved as children start school on children's mathematics learning, and the importance of providing meaningful, challenging, and relevant mathematical experiences throughout the transition to school
- the clear entitlement of children and educators to have assessment and instructional pedagogies match the strengths of the learners and the teachers
- the importance that the aspirations of children, families, communities, educators, and educational organisations be recognised as legitimate and key determinants of actions, experiences, and successes in both transition to school and mathematics learning
- the overriding belief that young children are powerful mathematics learners and that they can demonstrate this power as they start school.

This chapter aims to provide insight into the multi-faceted transitions processes with which children, families, and educators engage, and to explore the implications of transitions to school for children's mathematics education.

Transition to school

Transition to school refers to the movement of a child from their prior-to-school setting to primary school.

In general terms, **transition to school** is usually taken to mean the movement of a child from their prior-to-school setting (be it home, childcare centre, preschool, other family-based care, or any combination of these) to their primary school setting. However, transition to school is a contested term in the research literature, with different definitions focusing on different aspects of transitions. For example, Perry et al. (2015) observe that the term 'transition to school' may incorporate any of the following:

- children's 'readiness' for school and 'adjustment' to the new school context
- processes of movement from one context to another
- the changing roles of children, families, and educators in their educational communities
- 'border-crossing' and 'rites of passage'
- continuity and change
- the time period in which the transition is occurring.

What do you remember about your own transition to school? Did you feel supported, and feel that the knowledge you brought with you to school was valued?

Opinion Piece 10.1

PROFESSOR SUE DOCKETT

Learn more
Go online to hear this Opinion piece being read.

Many children know a great deal of mathematics before they start school. For example, they have a great sense of what is 'fair' when it comes to dividing up goods, they can often support an argument about why they should have the next turn, and they can recognise a range of numbers and shapes. This knowledge is important—but equally important is each child's disposition to use and build on that knowledge. Dispositions are 'habits of mind' that guide our actions—helping us use what we know to act in some way.

Much of my work about transition to school acknowledges the importance of dispositions and the ways that educators can promote dispositions such as confidence, perseverance, creativity, and playfulness. I am particularly interested in the relationships built around the time of transitions—between and among children, educators, families, and communities—and the opportunities these provide for all to recognise what children bring to school, to value this, and to extend it so that children not only gain knowledge, but also a willingness to use that knowledge in their interactions. Some of the most effective educators I have met have not only encouraged dispositions like persistence and creativity, they have also demonstrated them themselves.

Sue Dockett is Emeritus Professor of Early Childhood Education at Charles Sturt University, Albury, Australia. Over more than thirty years, she has been actively involved in early childhood education as a teacher, academic, and researcher. Sue has undertaken extensive research around educational transitions; in particular, transitions to school and the expectations, experiences, and perceptions of all involved. Sue has produced a wide range of research publications—co-authored with Bob Perry—which have had a substantial impact on policy, practice, and research. Sue is currently engaged in national and international research projects that aim to promote a positive start to school for all children.

In 2010, a group of Australian and international transition to school researchers came together to develop a common position on what they believed was important about transition to school (Perry et al., 2015). The result was the *Transition to School Position Statement* (Educational Transitions and Change Research Group [ETC], 2011). The Position Statement describes transition to school as being characterised by four 'pillars': *opportunities*, *aspirations*, *expectations*, and *entitlements*. In their 2015 text, Perry et al. present a commentary on what these four pillars might mean for *mathematics* and transitions to school. This commentary is summarised here.

OPPORTUNITIES

Opportunities are the chances and possibilities available to children as they start school.

Opportunities are the chances and possibilities available to children as they start school. Young children know a great deal of mathematics, and starting school presents wonderful opportunities for teachers and families to build on what children know and how they know it (Perry et al., 2015):

> Clearly, there are many opportunities afforded for enhanced mathematical experiences as children start school. Mathematical experiences in the home, in prior-to-school settings and in schools also afford opportunities for educators and family members to come together, explore and discuss children's learning. By building on the strengths of all involved in the child's transition to school, opportunities abound to enhance the experience for all (p. 4).

ASPIRATIONS

Aspirations are the positive hopes for children as they start school.

Everyone hopes the very best for children as they start school. These positive hopes can be described as our **aspirations** for children. Educators aspire to form strong relationships with children and families, parents want to be part of their children's mathematics education, and children hope that they will learn all kinds of new things as they start school (Perry et al., 2015):

> Adults—both educators and parents—and systems all have high aspirations for the mathematics learning of 'their' children ... Children also have aspirations or hopes for their own learning as they start school (Dockett & Perry, 2007) ... [including] enhanced mathematics learning experiences as they get older and move into school (pp. 5–6).

EXPECTATIONS

Expectations are ideas about what school will be like, and what it means to be a school student.

Expectations refers to ideas about what school will be like, and what it means to be a school student (ETC, 2011). Our expectations for children can affect their behaviour and achievement; and curriculum expectations, indeed, can greatly impact children and educators. There is often a mismatch between the planned mathematics curriculum and children's mathematical ability. Young children starting school may encounter a great deal of mathematics that they already know but which the teacher is determined to 'teach' them (Perry et al., 2015):

> [We] argue for educators to expect that children will start school knowing some mathematics and to recognise children's strengths in both knowing and learning (p. 6).

Families also have expectations as their children start school. They expect that they will play a part in their children's education, including their mathematics education. Partnerships between educators and families allow expectations to be shared (Perry et al., 2015):

[Families] expect that they will be respected and that their knowledge of their children will be listened to and acted upon … Families expect that their children's strengths, including those in mathematics, will be recognised and developed (p. 7).

ENTITLEMENTS

Entitlements are the rights that children have as they start school. Young children are entitled to a high-quality mathematics education as they transition to school. One way is through the provision of appropriate curricula and teacher expertise. This means that educators, too, have entitlements in the transition to school. They are entitled to the resources to do their job effectively, and to the respect of children, families, and systems (Perry et al., 2015):

Entitlements are the rights that children have as they start school.

They need time to listen to the children, to talk with the children, and to assess the children's mathematics learning in ways that are meaningful to all concerned, so that they know from where to start their teaching (p. 8).

Families also have entitlements as their children start school. They are entitled to know that their children are safe and valued. They also have the right to be respected as partners in their children's education (Perry et al., 2015).

Principles of effective transitions practices

As a result of their work on the *Starting School Research Project*, Dockett and Perry (2006) developed a series of *Guidelines for Effective Transition to School Programs*. These guidelines are underpinned by ten principles that support effective transitions practices. The ten principles are presented in Table 10.1.

TABLE 10.1 Principles of effective transition programs

PRINCIPLE 1	Effective transition programs establish relationships between the children, parents, and educators.
PRINCIPLE 2	Effective transition programs facilitate each child's development as a capable learner.
PRINCIPLE 3	Effective transition programs differentiate between 'orientation to school' and 'transition to school' programs.
PRINCIPLE 4	Effective transition programs draw upon dedicated funding and resources.
PRINCIPLE 5	Effective transition programs involve a range of stakeholders.
PRINCIPLE 6	Effective transition programs are well planned and effectively evaluated.
PRINCIPLE 7	Effective transition programs are flexible and responsive.
PRINCIPLE 8	Effective transition programs are based on mutual trust and respect.
PRINCIPLE 9	Effective transition programs rely on reciprocal communication among participants.
PRINCIPLE 10	Effective transition programs take into account contextual aspects of community, and of individual families and children within that community.

(Adapted from Dockett & Perry, 2006)

These principles can be used to guide transitions practices to support children's mathematics education as they transition to school. Tables 10.2 – 10.11 presents a series of questions adapted from Dockett and Perry (2006) to help you, as educators, think about how transition to school practices can recognise and support children's mathematical competence as they start school.

Pedagogies of educational transitions

'When children start school, there is a lot more going on for them than just their mathematics learning' (Perry & Dockett, 2005a, p. 66). A particular challenge is that transition to school 'is often characterised by discontinuity across the areas of relationships, pedagogy, curriculum, resources and support' (Dockett & Perry, 2014, p. 8). Perry and Dockett (2005a) elaborate on the dislocation that may characterise transition:

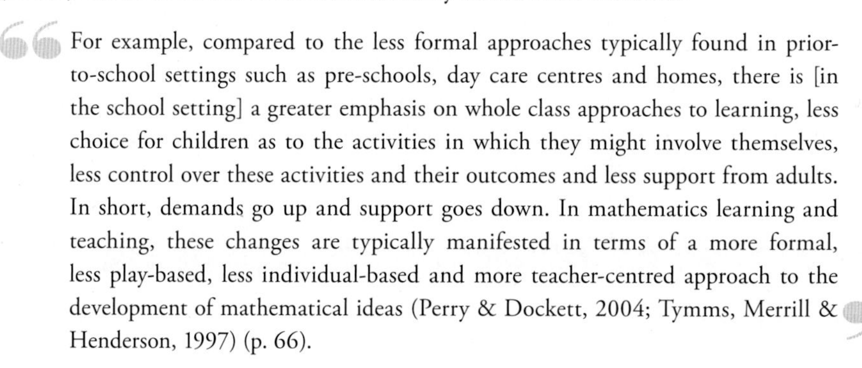

> For example, compared to the less formal approaches typically found in prior-to-school settings such as pre-schools, day care centres and homes, there is [in the school setting] a greater emphasis on whole class approaches to learning, less choice for children as to the activities in which they might involve themselves, less control over these activities and their outcomes and less support from adults. In short, demands go up and support goes down. In mathematics learning and teaching, these changes are typically manifested in terms of a more formal, less play-based, less individual-based and more teacher-centred approach to the development of mathematical ideas (Perry & Dockett, 2004; Tymms, Merrill & Henderson, 1997) (p. 66).

Tables 10.2 – 10.11 lay out the principles and questions to guide mathematics and transitions to school (adapted from Dockett & Perry, 2006).

TABLE 10.2 Principle 1 of mathematics and transitions to school

PRINCIPLE 1	QUESTIONS TO GUIDE MATHEMATICS EDUCATION PRACTICES
Effective transition programs establish positive relationships between the children, parents, and educators	» What connections are made between families and the school prior to children starting school? What information about children's mathematics learning is shared between families and the school? » What connections are made between educators in prior-to-school settings and educators in school? What information about children's mathematics learning is shared among educators? How do these groups communicate and collaborate around children's mathematics? » How do families find out about mathematics at school? » How is an educational partnership to support children's mathematics promoted?

TABLE 10.3 Principle 2 of mathematics and transitions to school

PRINCIPLE 2	QUESTIONS TO GUIDE MATHEMATICS EDUCATION PRACTICES
Effective transition programs facilitate each child's development as a capable learner	» How do educators in schools find out about children's mathematical abilities and interests? » Do first-year-of-school teachers visit children in prior-to-school settings? Do they see and explore the mathematics with which children engage in those settings? » Do prior-to-school educators visit schools? Do they see and explore the mathematics with which children engage in the school setting? » What documentation of children's mathematical strengths is shared between families, prior-to-school services, and schools? Are processes in place to ensure that this is appropriate documentation of children's mathematical knowledge, used in professional and ethical ways? » Is the format for collecting information about children's mathematical knowledge centred on identifying their strengths? » What information about children's mathematics is sought by schools from children and families, or prior-to-school services? How is this information used? » What meaningful opportunities are provided for children to demonstrate their mathematical competencies?

TABLE 10.4 Principle 3 of mathematics and transitions to school

PRINCIPLE 3	QUESTIONS TO GUIDE MATHEMATICS EDUCATION PRACTICES
Effective transition programs differentiate between 'orientation to school' and 'transition to school' programs	» Are there opportunities for two-way interaction about children's mathematics education or is the focus on school staff telling families what will occur, or school educators telling prior-to-school educators what children need to know about mathematics before they start school?

TABLE 10.5 Principle 4 of mathematics and transitions to school

PRINCIPLE 4	QUESTIONS TO GUIDE MATHEMATICS EDUCATION PRACTICES
Effective transition programs draw upon dedicated funding and resources	» What funding and/or resources are available for mathematics education programs? » How are the resources of prior-to-school and school settings used in an integrated way to support children's mathematics education? » How are community resources utilised within mathematics education programs?

TABLE 10.6 Principle 5 of mathematics and transitions to school

PRINCIPLE 5	QUESTIONS TO GUIDE MATHEMATICS EDUCATION PRACTICES
Effective transition programs involve a range of stakeholders	» Who is involved in the team responsible for the planning and implementation of mathematics education programs? Whose knowledge and views of mathematics education are represented? » How do children have input into programs? How do they share what is important to them about their mathematics education? » How do families have input into programs? How do they share what is important to them about their children's mathematics education?

TABLE 10.7 Principle 6 of mathematics and transitions to school

PRINCIPLE 6	QUESTIONS TO GUIDE MATHEMATICS EDUCATION PRACTICES
Effective transition programs are well planned and effectively evaluated	» What provisions are made for the planning and evaluation of programs? » What provisions exist for input from families, children, and educators in prior-to-school settings? » What data are collected about children's mathematics education, and from whom? How can this be used to inform future programs? » How is data about children's mathematics education reported to children, families, educators, and the community?

TABLE 10.8 Principle 7 of mathematics and transitions to school

PRINCIPLE 7	QUESTIONS TO GUIDE MATHEMATICS EDUCATION PRACTICES
Effective transition programs are flexible and responsive	» How do staff in these programs respond to requests for information or advice about children's mathematics? » How do programs respond to the mathematical needs or interests of the participants?

TABLE 10.9 Principle 8 of mathematics and transitions to school

PRINCIPLE 8	QUESTIONS TO GUIDE MATHEMATICS EDUCATION PRACTICES
Effective transition programs are based on mutual trust and respect	» What opportunities are there for listening to children, families, prior-to-school educators, and staff in schools about children's mathematics education? » How is a climate of trust and respect about children's mathematics education generated and maintained? » Is cultural diversity understood and respected in relation to mathematics education?

TABLE 10.10 Principle 9 of mathematics and transitions to school

PRINCIPLE 9	QUESTIONS TO GUIDE MATHEMATICS EDUCATION PRACTICES
Effective transition programs rely on reciprocal communication among participants	» What do those involved in transition programs regard as reciprocal communication about children's mathematics? » How is reciprocal communication about children's mathematics encouraged? » What opportunities exist for educators in prior-to-school and school settings to consider mathematics and transition to school from other perspectives—for example, that of a child, parent, or community member? » How does communication within programs indicate that the perspectives about mathematics of all contributors are valued? » What policies and procedures are in place to ensure that communication about children's mathematics is both ethical and professional?

TABLE 10.11 Principle 10 of mathematics and transitions to school

PRINCIPLE 10	QUESTIONS TO GUIDE MATHEMATICS EDUCATION PRACTICES
Effective transition programs take into account contextual aspects of community, and of individual families and children within that community	» How do those involved in programs ensure that differences among children's and families' mathematical knowledge are not perceived as deficits? » What processes and procedures ensure a focus on children's and families' mathematical strengths, rather than their limitations? » How are specific aspects of the mathematics in local communities reflected in transition programs? » What professional development opportunities are available for educators (in both prior-to-school and school settings) to get to know more about working with children and families with diverse mathematical backgrounds and needs? » What changes can be made to programs to make sure that all families and children feel included in mathematics education?

As MacDonald et al. (2016) highlight, 'the move from so-called "play-based and child-centred" pedagogies in prior-to-school settings to more formal, "subject-based" pedagogies in schools would seem to challenge the often heard call for continuity across the transition to school' (p. 167). However, 'there does seem to be agreement that promoting continuity does not mean that contexts should become the same … Indeed, there is strong evidence that young children want school and prior-to-school to be quite different; they do not want more of the same as they start school' (Perry, Dockett & Harley, 2012, p. 157).

One way of promoting *continuity*, while also allowing for and encouraging *change*, is through **pedagogies of educational transitions**. Such pedagogies help to establish connections across the educational settings. Educators in prior-to-school settings can support children's transition to school mathematics by acknowledging the mathematical language, curricula, and pedagogical approaches utilised in school settings and making links to these through meaningful documentation of, and communication about, mathematics. Equally, school educators can support children's transitions by valuing and utilising pedagogical approaches common to prior-to-school settings as a way of helping children feel comfortable in the school environment and feel that their mathematical knowledge and ways of learning are valued in their new educational context.

Pedagogies of educational transitions help to establish connections across educational settings.

Opinion Piece 10.2

DR JESSAMY DAVIES

As part of my doctoral research, it was important to develop a working definition of what I meant by 'pedagogy', and how that was relevant for educational transitions. The definition of pedagogies of educational transition that I developed was: 'The interactive processes and strategies that enable the development of opportunities, aspirations, expectations, and entitlements for children, families, educators, communities, and educational systems around transition to school, together with the theories, beliefs, policies, and controversies that shape them'.

One key element of this definition is that, collectively, the 'processes and strategies' that are used by educators form their pedagogical practice. This refers to those processes and strategies that may be conscious, planned, and deliberate, but also those that are not. Another important element refers to the impact of the *influences* on those processes and strategies, and the acknowledgement that these are constantly shaping the pedagogies of educators.

It is important to note that effective pedagogies that enhance continuity and help promote effective transition to school do not occur without several important processes. In my research, educators that worked collaboratively through educator networks were able to develop positive relationships with one another, share common understandings, communicate effectively, and hold professional respect for one another. In turn, this enabled positive pedagogical practices to be implemented to enhance the transition to school experiences for all involved.

Jessamy Davies is a Lecturer in Education at Charles Sturt University, Albury–Wodonga, Australia. Her doctoral work was undertaken as part of an Australian Research Council Discovery project that explored policy–practice trajectories at the time of transition to school. The project examined the policy intentions and impact of the *Early Years Learning Framework* and the *Australian Curriculum* on transition to school at national, state, and local levels.

Learn more
Go online to hear this Opinion Piece being read.

Communicating across the transition to school

Communication between prior-to-school and school educators is vital for supporting children's mathematics education as they transition to school. This principle of communication and information-sharing is reflected in the EYLF (DET, 2019), which states:

> As children make transitions to new settings (including school) educators from early childhood settings and schools commit to sharing information about each child's knowledge and skills so learning can build on foundations of earlier learning. Educators work collaboratively with each child's new educator and other professionals to ensure a successful transition (p. 19).

Of course, communication between educators is not the only form of communication that is essential for supporting children's transition from prior-to-school to school settings. Parents and other family members have a great deal of knowledge about their children's learning—including their mathematics learning—and this information should be sought and valued during the transition to school.

Perhaps most important of all is communication with the children themselves. Children are certainly capable of communicating their own mathematical knowledge to their educators as they transition to school. There are a number of ways in which children might communicate their mathematical knowledge; but a good starting point is to simply talk with the child. As Perry and Dockett (2005b) explain:

> Sitting and listening to young children, even for a short time, can provide great insight into the power of these children in all sorts of areas, including mathematics. It behoves all early childhood educators to do this with all of their children so that their current learning is recognised and used and so that our young children are challenged by their mathematics learning and find that mathematics can be an exciting subject. There is ample evidence in both prior-to-school and school settings that this can be done. The aim is to ensure that all children experience quality practice in all of their educational settings (p. 36).

Clearly, it is important for all stakeholders in the transition process—children, families, and educators—to share information that supports the child's learning as they move from one setting to another.

TRANSITION STATEMENTS

Transition statements are summaries of children's prior-to-school learning.

Transition statements are summaries of children's prior-to-school learning, used in many parts of Australia as a means of communicating children's learning from their prior-to-school setting to their school setting. As the New South Wales Department of Education and Communities (NSW DEC, 2017) explains, a transition statement

> records a child's interests, strengths and preferred ways of learning in their year prior to school. Its purpose is to assist early childhood educators, parents/carers and primary school teachers to better understand a child and how best to support their transition from early childhood education to school (p. 1).

FIGURE 10.1 NSW Transition to School Statement pp. 2–3

SECTION B. Early childhood educator to complete this section

Section B is guided by the *Early Years Learning Framework* and aligns with the five Learning Outcomes. When **sections B** and **C** are completed, please take a copy of it and pass the whole Transition to School document to the child's parents/carers for them to complete **sections D, E** and **F**.

1. **Briefly summarise your professional views on this child's independence and resilience.**
   (Outcome 1: Children's Identity)

2. **Briefly summarise how you see this child builds relationships with peers end adults.**
   (Outcome 2: Children's Connection and Contribution to the World)

3. **Briefly summarise how this child self-regulates and manages their emotions.**
   (Outcome 3. Children's Wellbeing)

4. Briefly summarise how curious this child is to learn now things and their ability to persist at tasks.
(Outcome 4: Children as Confident Learning)

5. Briefly summarise how you see this child's communication skills, taking into account language and literacy.
(Outcome 5: Children as Effective Communicators)

6. What are the child's overall strengths?

7. What are some of the child's interests?

(Source: NSW DEC, 2017)

Generally speaking, transition statements provide information designed to make a child's passage to school as seamless as possible. For example, the Victorian Department of Education and Training (VIC DET, 2017) requires that such a statement should:

- summarise a child's learning and development as they enter school
- identify their individual approaches to learning and interests
- indicate how the child can be supported to continue learning.

Transition statements are completed by the family and the early childhood educator and include sections that seek the child's views and input (for example, a picture they have drawn). The statements are linked to the EYLF and encourage communication about the child's learning and development. As such, they present a great opportunity for educators and families to share information about a child's mathematical learning. For example, the NSW Transition to School Statement (Figure 10.1; NSW DEC, 2017) asks educators, among other things, to summarise how curious the child is to learn new things and their ability to persist at tasks (Outcome 4: Children as Confident Learners); the child's overall strengths; and some of the child's interests—all of which present an opportunity to talk about a child's mathematical interests and understandings. The Queensland Transition Statement (Queensland Department of Education and Training [QLD DET], 2013) similarly requires educators to comment on each of the EYLF outcomes, and an example of how mathematical learning can be included in such comments is shown in Figure 10.2.

How confident do you feel that you could include information about a child's mathematical learning in a transition statement? How might the EYLF and the *Australian Curriculum: Mathematics* documents assist you in this task?

Learning outcome: Children are effective communicators

FIGURE 10.2 Excerpt from example Queensland Transition Statement

A kindergarten child who is an effective communicator:	Connor:
» explores and expands ways to use language » explores and engages with literacy in personally meaningful ways » explores and engages with numeracy in personally meaningful ways.	» is a confident communicator who enjoys sharing ideas and posing questions to the group for investigation » is an attentive listener and uses the social conventions for turn-taking most times » loves reading nonfiction texts about wildlife and will intentionally seek out books related to a topic he is interested in. He is demonstrating a strong understanding of reading for a purpose. » is using combinations of letters to create signs for his play, recognises a few sounds and likes to engage in rhyming word play games. » incorporates numerals and letters in play—for example, attempts to copy letters from the computer onto the pretend laptop he made and drawing numerals on the pretend mobile phone he made. » uses mathematical language in daily routines—for example, uses language related to time to talk about what's happening in the day ('before', 'after', 'later', 'next'), has one-to-one correspondence to 20, recognises numerals 1–10, and can create a two-part pattern.

(Source: QLD Department of Education and Training, 2013)

Chapter summary

Transition to school is a crucial time in a young child's life. It is important for educators to support children's transitions in ways that help children to know that their knowledge—including their mathematical knowledge—is recognised and valued in both their prior-to-school and school settings. Educators can use pedagogies of educational transitions to help support children's mathematical learning across the transition to school. Attention to the principles of effective transitions practices can help educators to recognise, communicate, and celebrate children's mathematical knowledge as they start school.

The specific topics we covered in this chapter were:

- transition to school
- opportunities
- aspirations
- expectations
- entitlements
- principles of effective transitions practices
- pedagogies of educational transitions
- communicating across the transition to school
- transition statements.

FOR FURTHER REFLECTION

All stakeholders in the transition to school process (children, families, and educators) have a role to play in communicating children's mathematical learning from their prior-to-school setting to their school setting.

1. In what ways could you encourage prior-to-school and school educators to communicate about children's mathematics learning?
2. In what ways could you encourage parents to communicate with prior-to-school and school educators about their child's mathematics learning?
3. In what ways could you encourage children to communicate their own mathematics learning to prior-to-school and school educators?

FURTHER READING

Educational Transitions and Change (ETC) Research Group. (2011). *Transition to school: Position statement*. Albury–Wodonga, Australia: Research Institute for Professional Practice, Learning and Education (RIPPLE), Charles Sturt University. Available online: https://arts-ed.csu.edu.au/education/transitions/publications/Position-Statement.pdf

Perry, B., Dockett, S., & Harley, E. (2012). The Early Years Learning Framework for Australia and the Australian Curriculum: Mathematics—Linking educators' practice through pedagogical inquiry questions. In B. Atweh, M. Goos, R. Jorgensen, & D. Siemon (eds), *Engaging the Australian Curriculum Mathematics: Perspectives from the field* (pp. 153–74). Adelaide: MERGA.

Perry, B., MacDonald, A., & Gervasoni, A. (2015). Mathematics and transition to school: Theoretical frameworks and practical implications. In B. Perry, A. MacDonald, & A. Gervasoni (eds), *Mathematics and transition to school: International perspectives* (pp. 1–12). Singapore: Springer.

11 Assessment and Planning

Learn more
Go online to see a video from Paige Lee explaining the key messages of this chapter.

CHAPTER OVERVIEW

This chapter will explore the reciprocal relationship between assessment and planning, and the different ways of influencing children's opportunities to learn about mathematics. The chapter also introduces the use of Learning Experience Plans for assessing, documenting, and planning for children's mathematical learning. You will notice that Learning Experience Plans appear as a pedagogical feature throughout the subsequent chapters in this book as a means of showing you how these plans can be used in different ways to provide mathematical learning opportunities for children.

In this chapter, you will learn about the following topics:

- assessment and planning
- ELPSARA framework
- documenting and communicating mathematical learning
- photographs
- conversations
- documentation panels
- drawings
- learning stories
- observing and interpreting mathematical learning
- planning for mathematical learning
- Learning Experience Plans.

KEY TERMS

- Assessment
- Documentation panels
- Learning Experience Plans
- Learning stories
- Planning

AMY MACDONALD

Introduction

Assessment can be defined as the process of collecting, organising, and analysing information about children's performance.

This chapter looks at the relationship between assessment and planning in early childhood mathematics education. **Assessment** can be defined as the process of collecting, organising, and analysing information about children's performance (Perry & Conroy, 1994). Assessment features prominently in the *Early Years Learning Framework for Australia* (EYLF; DET, 2019), with a clear articulation of the purposes of, and approaches to, assessment that should be considered in the early childhood years. The EYLF states that:

> Educators use a variety of strategies to collect, document, organise, synthesise and interpret the information that they gather to assess children's learning. They search for appropriate ways to collect rich and meaningful information that depicts children's learning in context, describes their progress and identifies their strengths, skills and understandings (p. 20).

Furthermore:

> More recent approaches to assessment also examine the learning strategies that children use and reflect ways in which learning is co-constructed through interactions between the educator and each child. Used effectively, these approaches to assessment become powerful ways to make the process of learning visible to children and their families, educators and other professionals (p. 20).

These views are also reflected by the Australian Association of Mathematics Teachers (AAMT; 2006), which states:

> Early childhood educators should assess young children's mathematical development through means such as observations, learning stories, discussions, etc. that are sensitive to the general development of the children, their mathematical development, their cultural and linguistic backgrounds, and the nature of mathematics as an investigative, problem solving and sustained endeavour.

Planning refers to the process of using information to inform the provision of learning experiences.

Taking such a stance in relation to *assessment* also influences the ways in which we view *planning*. In this text, **planning** refers to the process of using information gathered from children, families, and other sources to inform the provision of mathematical learning experiences and opportunities. There is a reciprocal relationship between assessment and planning; planning is based on, and informed by, assessment of children's prior experiences and understandings, and in turn, the experiences that are planned are then assessed in sensitive and responsive ways.

Assessment and planning

In the context of early childhood education, it is important to think about assessment and planning as two interrelated processes that inform one another. While it is of course important to understand assessment approaches and planning approaches in their own right, it is much more powerful to consider how one can be used to enhance the other. To emphasise this relationship, assessment and planning should be thought of as a continuous cycle that informs children's learning in mathematics education.

ELPSARA FRAMEWORK

A useful model for thinking about the planning and assessment process in mathematics education is the 'ELPSARA'—an acronym that stands for the following elements:

1 E is for Experience.
2 L is for Language.
3 P is for Pictorial Representation.
4 S is for Symbolic Representation.
5 A is for Application.
6 R is for Reflection.
7 A is for Assessment.

A visual representation of the ELPSARA framework is presented in Figure 11.1. The ELPSARA framework was originally adapted from the work of Liebeck (1984), but has undergone many developments during the years since, particularly as a result of the work of academic Tracey Smith with teacher education students at Charles Sturt University (Meaney & Lange, 2011). The different aspects of the ELPSARA framework can be best understood by thinking that *if* mathematical learning can be enhanced in certain ways, *then* there are implications for the planning of mathematical learning experiences (Meaney & Lange, 2011). Table 11.1 explains some of these links.

FIGURE 11.1 ELPSARA framework

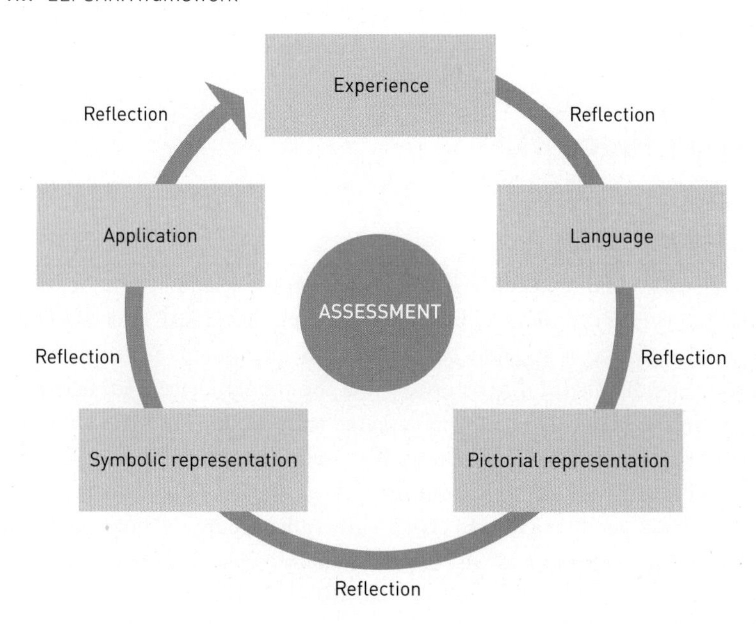

(Adapted from Meaney & Lange, 2011)

TABLE 11.1 Links between mathematical learning and teaching using ELPSARA

IF LEARNING IS ENHANCED THROUGH:	*THEN* THE IMPLICATIONS FOR TEACHING MIGHT BE:
EXPERIENCES—making prior experiences explicit and being able to connect these experiences to new knowledge and experiences.	Find out about children's prior experiences and knowledge *before* teaching a concept or topic so connections can be made explicit.
LANGUAGE—actively refining the particular language of mathematics to make a personal sense of it.	Encourage talking and writing to make new language explicit and to develop a shared understanding of children's current language development.
PICTORIAL REPRESENTATION—experiencing mathematical concepts in pictorial or concrete form.	Provide opportunities for representing concepts in pictorial and concrete form and interpreting pictures and models that represent concepts.
SYMBOLIC REPRESENTATION—being able to make the transition from pictures to symbols.	Decide when and how to introduce and scaffold more abstract or symbolic representations into children's experiences *and* recognise when children are already at this level of understanding.
APPLICATION OF KNOWLEDGE—understanding the significance of learning by applying new knowledge to solve problems in meaningful contexts that represent life experiences.	Design learning experiences that are relevant and meaningful to children's life experiences.
REFLECTION—a calm, lengthy, intentional consideration.	By having children reflect on their strategies, their own thinking processes will strengthen as thinking becomes a learning experience in itself.
CONSTRUCTIVE ASSESSMENT—assessment practices that inform and guide planning and teaching and provide evidence of what children know, understand, and can do in mathematics.	Constructive forms of assessment are essentially social processes taking place naturally in social classroom settings as an integral part of everyday learning and teaching.

(Adapted from Meaney & Lange, 2011)

Opinion Piece 11.1

Learn more
Go online to hear this Opinion Piece being read.

MICHELLE MULLER

Within the early childhood education and care sector in which I work, we use various methods of assessment to determine children's mathematical understandings. So much of what we do is spontaneous, while also intentional. When we are eating morning tea we discuss the different fruits that the children have, and we might make a graph using the interactive whiteboard, and note the number of apples, bananas, and mandarins that the children might have. We discuss size, 'Goodness that is a very big apple!!', and quantity, 'Now let's count the tally of oranges to see how many we have altogether'. Incidentally, we start to classify into other groups, identifying which ones have skin that we can eat, and which ones we need to peel before we eat. From this we build upon the children's knowledge of different types of fruit, size, colour, etc. and the ways that such attributes can be grouped together. We use this information to provide other opportunities for children either to consolidate their understandings or to establish them, and we know that different children develop such understandings in very different ways. Recently, I watched a small group of children as they classified a basket of plastic animals. I had intentionally provided a mixture of sea and land animals. One child placed each of the tigers (about eight) on one long line, and then moved them around, putting the largest at one end of the line, and the shortest at the other. She then placed elephants together, giraffes together, and the sea creatures in a different basket. I heard her say to the other children (who were also moving the animals into groups), 'I can't put the stingray with the others—they don't live together'. From this, we can make an assumption that this child has used classification methods to sort the animals. The child demonstrates her emerging mathematical skills through her use of seriation, comparing, and grouping, as well as her use of her knowledge of animals. My appreciation of her skills provides me with a place to go to enhance her learning even further.

As professionals, we are constantly looking for ways to embed maths learning into our space. It might be snipping play dough into pieces and then counting the pieces, noting one-to-one correspondence, counting on in order, or it could be when we bake with the children and discuss measurement—we use language such as 'one half', 'two thirds', etc. and show the quantity attached to the amount. By embedding maths learning, and maths language into the program, we are able spontaneously to assess the children's knowledge, and in turn, we are provided with opportunities to extend a child's (and children's) mathematical awareness.

Michelle Muller is an experienced early childhood educator and centre director, and a sessional academic at Charles Sturt University, Albury–Wodonga, Australia. She has worked for a number of years in both private and community-based services. Michelle works with other educators in providing a rich program that supports the diversity of children and families, and is a strong advocate for the rights of children. Michelle has a keen interest in leadership in early childhood and is a member of a number of strong network groups in the Albury–Wodonga area.

How might you involve children and families in the assessment and planning process?

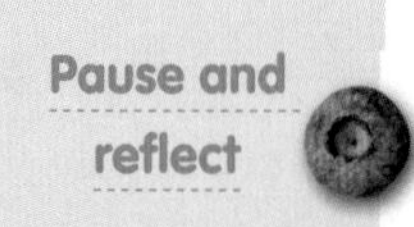

Documenting and communicating mathematical learning

Assessment in mathematics education can be considered as 'the process of gathering evidence about a student's knowledge of, ability to use, and disposition toward, mathematics and of making inferences from that evidence for a variety of purposes' (National Council of Teachers of Mathematics, 1995, p. 3). Put another way, assessment is about 'observing, recording, and otherwise documenting the work that children do and how they do it' (Copley, 1999, p. 183). As Copley explains:

> early childhood teachers and parents have often instinctively used effective assessment practices as they attempt to understand children and their experiences. Detailed observations, probing questions, and frequent listening sessions are all part of adults' repertoires as they try to understand what a child knows, how he learns, and what to teach next (p. 182).

Assessment is essentially about making children's learning visible, and there are a number of ways in which this might occur. The way you approach assessment will depend on you, your philosophies, the context in which you work, and—most importantly—the children with whom you work. Aitken, Hunt, Roy, and Sajfar (2012, pp. 28–9) highlight approaches that are appropriate for the early childhood years, which include:

- photographs
- conversations
- documentation panels
- drawings
- learning stories.

PHOTOGRAPHS

Selective photographs that capture children as they are engaged in learning can be valuable records (Aitken et al., 2012). Children can participate in taking the photographs and choosing what to photograph. Children can also use cameras and tablets to record their own mathematical learning and create their own mathematical learning stories (Arthur et al., 2015).

CONVERSATIONS

A conversation with a child about their mathematical explorations can provide in-depth evidence of learning (Aitken et al., 2012). Open-ended questioning during conversations invites the child to share their ideas and understandings (Figure 11.2). When recorded, transcripts of conversations can provide many insights into children's mathematical thinking and uses of mathematical language (Arthur et al., 2015).

FIGURE 11.2 Conversations with children about their mathematical explorations

(Source: Michelle Muller)

DOCUMENTATION PANELS

Educators can collect work samples, recorded conversations, photographs, and anecdotal evidence, and display these as a **documentation panel** (Figure 11.3). The panel should be placed where the children can see it, and children should be invited to share their learning with their families or other visitors. Questions that focus on mathematical learning can be added to encourage discussion.

Documentation panels are displays of work samples, recorded conversations, photographs, and anecdotal evidence.

FIGURE 11.3 Documentation panel

(Source: Maree Parkes/Amy MacDonald)

DRAWINGS

A child's drawing can be a useful source of information about their mathematical learning. Drawings become more powerful forms of assessment when the child's comments are added to the drawing and the drawing and comments are considered as a whole. When drawings are collected and dated, they can provide useful insights into the child's mathematical development over time (Arthur et al., 2015).

As indicated in Chapter 4, the 'draw a clock' task is an open-ended drawing task I have used many times with very young children both as part of my teaching and as part of my research (see, for example, MacDonald, 2013; Smith & MacDonald, 2009). This drawing task provides a useful method of assessing children's understandings about time, and the measurement and representation of time. Children are given a blank piece of paper and are simply asked to 'draw a clock'—no further direction is given. This drawing activity gives children the opportunity to share what they know about clocks and the representation of time, including things like sequencing numbers and unit iteration. Some children also take the opportunity to record different 'times' that they know on the clock. Children may choose to draw a range of different clocks, including both digital and analogue clocks, watches, cuckoo clocks, alarms clocks, and so forth. Many children also include an accompanying 'narrative', which reveals their different experiences with clocks in a range of contexts and gives insight as to how their understandings have developed (Figure 11.4).

FIGURE 11.4 Nathan's drawing of clocks

(Source: Amy MacDonald)

Other drawing prompts I have used to learn about children's mathematical experiences and understandings include:

- draw a ruler
- draw yourself measuring
- draw something tall and something short
- draw something heavy and something light
- draw something hot and something cold (MacDonald, 2013).

Pause and reflect

What other drawing prompts can you think of to encourage children to draw about their mathematical experiences and understandings?

LEARNING STORIES

Learning stories, pioneered by Margaret Carr in New Zealand for use with the *Te Whariki* early childhood curriculum, are qualitative snapshots, recorded as structured written narratives, often with accompanying photographs, that document and communicate the context and complexity of children's learning (Carr, 2001). Learning stories focus on children's strengths and are positioned within a social and cultural context (Arthur et al., 2015). Learning stories may differ in length, the amount of detail included, whether they focus on one child or a group, and their structure (Hunting, Mousley, & Perry, 2012). In general, the construction of learning stories involves three key phases: notice, reflect, and respond (Aitken et al., 2012). More specifically, mathematics-focused learning stories may include any or all of the following components:

- description of the context and what happened
- analysis of the mathematical learning
- links to curriculum frameworks such as the EYLF or the *Australian Curriculum*
- how the physical and/or social environment supported the learning
- reflection upon 'where to now' to build on the child's learning
- a note to the child to provide feedback on their mathematical development
- a note to the family to inform them as to their child's mathematical learning.

Importantly, a mathematics-focused learning story should *unpack* the mathematical learning that has taken place, attending closely to the mathematical concepts being developed. For example, rather than just saying 'Audrey was counting the blocks', educators should seek to explain the *concepts* involved in counting—for example, 'As Audrey counted the blocks, she demonstrated developing understandings of one-to-one correspondence, numeral names, and the stable order principle' (MacDonald, 2015a).

An example of a mathematics-focused learning story, written by early childhood educator Valerie Tillett, is presented in 'Where did I hide it?'.

Observing and interpreting children's mathematical learning

In order to construct learning stories or other forms of documentation that usefully communicate children's mathematical learning, we must develop skills in observing and interpreting children's mathematical learning (MacDonald, 2015a). Resources such as the *Early Childhood Learning* DVD published by DEEWR (2006) are helpful for developing these skills, as watching videos can allow us to practise 'noticing' mathematics and articulating this mathematics to others. These skills in noticing children's mathematics can then be utilised in our everyday observations of children's learning. It is only with careful observation that we can notice—and in turn, interpret—the potential for mathematical learning in children's activity (MacDonald, 2015a). A useful example of mathematical learning from the *Early Childhood Learning* DVD is the video 'Sorting the groceries'. Figure 11.5 shows a couple of stills from this video and Table 11.2 provides an example of how some of the observations taken from the video can be interpreted in relation to a child's mathematical learning.

FIGURE 11.5 Stills from the 'Sorting the groceries' video

(Source: Department of Education, Employment and Workplace Relations, 2006)

TABLE 11.2 Observing and interpreting mathematical learning

OBSERVATIONS	INTERPRETATIONS
Mum says, 'This one's so heavy, and this one's light'.	Measurement language (mass), comparison of masses, hefting (measuring by lifting)
Girl points to the bag and says, 'Got to put all the heavy stuff out'.	Measurement language, classification
Mum describes the weight of the items as she takes them out of the bag.	Measurement language, comparison of masses
Mum asks, 'Where does this one go?' Girl responds, 'In the pantry'.	Classification of items based on their characteristics (for example, dry items = pantry, frozen items = freezer)
Girl matches each item to its storage location.	Understandings of temperature—pantry is for dry items, fridge is for cold items, freezer is for frozen items
Girl finds a clear space in the freezer to put the ice cream.	Position, spatial awareness
Mum asks, 'Where does this one go?' Girl responds, 'Up the top'.	Positional language

(Source: MacDonald, 2015a, p. 49)

Planning for mathematical learning

It is important to think about planning as a reflective process. We need to be flexible, and continually think about how we can engage and sustain children's interest in the mathematical learning experiences that we provide. We must also be willing and able to incorporate children's ideas in this process—indeed, as the EYLF (DET, 2019) states, educators should be 'responsive to children's ideas and play, which form an important basis for curriculum decision making' (p. 17).

To guide a reflective planning process, Copley et al. (2007, pp. 80–1) provide the following questions for educators to consider:

WHERE DID I HIDE IT?

Zac, Henry, and Aidon were playing a hiding game—they would hide a small dinosaur from each other. It was Zac's turn to hide the dinosaur but after he did he forgot where he had hidden it. All three boys searched for it and could not find it. Henry commented to the boys, 'It's hard to find, just like treasure'.

The educator had heard their issue and asked the boys, 'It is hard to find something small in such a big playground. It reminds me of the pirate stories we listened to. Do you remember how the pirates knew how to find their treasure after they hid it?'

The boys looked puzzled at first. Then Henry said, 'A treasure map!'

'That's right,' said the educator. 'Maybe a map could help you to remember where you hid your dinosaur, just like the pirates' treasure map' (an interest area in pirates had been explored over several weeks prior).

'Yeah,' said Zac.

'My Mum has a map when she is driving,' said Henry.

'What does the map do?' asked the educator.

'It helps her find her work. It has the road on it and my house,' said Henry.

'Do you think you could make a map of the playground?' asked the educator.

'Yep,' said Henry.

'What do we need to make one?' asked the educator.

'Paper, big paper and pencils and we can draw on it,' said Henry.

'And pictures of where I hide it and stick it on the map,' added Zac.

They then discussed what the map would represent and how to capture the area of the playground that the map represented. The boys took photos of secret hiding places ready to stick on later in the day. The educator spoke to them about the space of the yard and how they could represent the area drawn on the paper. She started the map for them marking the corners of the playground and pointed out main structures to include in the map, talking about positioning/location of certain main structures. The boys took turns marking the sandpit, deck, and garden on the paper. This was followed by taking pictures of the hiding places using a camera. Henry commented, 'It will be easy to find the treasure with the map.'

After printing the photos, the boys cut them out and with the support of the educator they began to plot them onto the map representing the position/area where they had hidden their 'treasure'. During this collaboration, they discussed the position of the hiding spot in relation to the physical environment with each other and their educator in the actual environment.

The map was then used by the boys to hide their treasure (another dinosaur) and when they hid it they marked an 'X' on the picture to represent the position within the yard. The game continued as they boys took turns in hiding and locating the dinosaur using the map to assist them in locating the treasure.

Analysis of learning

Henry was able to relate and adapt his knowledge of how a map can locate an address/location as discussed with his mother to the purpose of the 'treasure map' for the game. His knowledge of, and confidence about, the concept of a map was then transferred to his peers and, with the support of the educator, the boys co-constructed the map for the purpose of their game. Using the map within the game provided a meaningful context, providing a concrete experience on which to build knowledge of location, space, positioning, mapping, distance, and plotting, as well as technological skills in using the digital camera to produce photos. Henry demonstrated that he was able to see similarities and connections to the hiding places (locations), with the educator's support. This learning experience provided an opportunity to transfer and adapt this knowledge about mapping, which had been introduced in the home.

Learning Outcome 4: Children are confident and involved learners

Henry demonstrated enthusiasm as a learner to further extend his interest in mapping, willing to explore technology and new strategies to solve the problem of how to communicate location while constructing a map. He confidently used feedback from the mathematical discussion with others to co-construct, revise, and build on ideas to construct the treasure map.

Learning Outcome 5: Children are effective communicators

Henry interacted with the other children to explore ideas, negotiate, and share his understandings of location and mapping. He was able to draw on, and confidently express, his understandings built on his prior experiences both at home and preschool to construct meaning using photographs and symbols to communicate location to construct the treasure map.

(Source: Valerie Tillett)

Before teaching, think about these questions:

- What do I want the children to know and be able to do?
- What do the children already know about this topic?
- What essential dispositions am I fostering?
- How will I evaluate and assess the children's learning?

During teaching, think about these questions:

- Is every child learning what I expected?
- Is unanticipated learning occurring?
- Are things going as planned?

After teaching, think about these questions:

- What worked? What is the evidence?
- What needs to be changed?
- What do I do next for the group as a whole? For individual children?

These questions should be considered when planning and implementing *any* mathematical learning experience, regardless of the pedagogical approach utilised.

Learning Experience Plans

A **Learning Experience Plan** is a structured planning document that articulates a learning focus, procedure, and review strategy for a learning experience. These may also be known as 'lesson plans' in primary school settings. The purpose of a Learning Experience Plan is to provide a scaffold for designing learning experiences that respond to, and extend, children's existing experiences and understandings in mathematics. While Learning Experience Plans may use different templates and have different features, they typically include the following information:

- *Experience information*—What is the name of the experience? For whom is it planned?
- *Learning focus*—What mathematical learning do you hope to facilitate? How does this relate to, and build on, children's strengths and interests? Why is this an appropriate learning experience for these children, at this time?
- *Learning outcomes*—What curriculum outcomes will the children work towards in this learning experience?
- *Requirements*—What materials are needed for this learning experience? Where, and how, will the experience be presented?
- *Procedure*—How will you implement the learning experience? What teaching strategies will you use? What will you say or do?
- *Plan for review*—How will you evaluate the learning experience? How will you know if the intended mathematical learning was facilitated? How will you judge the effectiveness of your strategies?

A **Learning Experience Plan** is a structured planning document that articulates a learning focus, procedure, and review strategy for a learning experience.

An example Learning Experience Plan prepared by early childhood educator Sussann Beer is presented in Learning Experience Plan 11.1. In this text, I advocate the use of Learning Experience Plans as a key tool for both assessment and planning in early childhood mathematics education. In the chapters that follow, example Learning Experience Plans appear as a pedagogical feature of the text that demonstrate the different ways in which these documents can be used to plan meaningful mathematical learning experiences for young children.

Learning Experience Plan 11.1

'Exploring the value of money'

Experience information

The experience is planned for 19 September 2016 with a group of five boys who are quite close as a group of friends and aged between four and five years old.

Learning focus

The key mathematical learning that I am hoping to facilitate is value, which is about applying the logic that things have value that can be measured, and that dollars and cents are measurement units. The children for whom I am planning this experience have strong ability in counting and recognising numbers. They are showing a current interest in money, using paper money, games that use money such as Monopoly and identifying different roles that handle money. Over the past couple of months of observing this group of boys during their last year of early childhood education and care, they have developed mathematical abilities and processes quite quickly and well. A few months ago the boys were able to identify numbers and count; however, now they are aware of the different uses of numbers and are able to use them and show an interest in different ways to use them. My knowledge of these children's development is that they are capable of using numbers and learning about the value of things and using different mathematical ideas.

I have chosen this learning experience for the group of boys because they have been role playing cashiers and identifying where, and for what purpose, money can be used. They have been identifying different types of money and asking about types of currencies around the world.

EYLF outcomes

Within Learning Outcome 4.2, 'Children develop a range of skills and processes such as problem solving, inquiry, experimentation, hypothesising, researching and investigating', the evidence I will be observing is 'children create and use representation to organise, record and communicate mathematical ideas and concepts', 'children make predictions, generalisations about their daily activities, using patterns they generate or identify and communicate these using mathematical language and symbols', and 'children contribute constructively to mathematical discussions and arguments'.

Within Learning Outcome 5.1, 'Children interact verbally and non-verbally with others for a range of purposes', the evidence I will be observing is 'children demonstrate an increasing understanding of measurement and number using vocabulary to describe size, length, volume, capacity and names of numbers'—although my experience is based on value and the value of money, I am expecting the children to be able to meet this outcome and be able to name the numbers on money that they see.

Within Learning Outcome 5.5, 'Children use information and communication technologies to access information, investigate ideas and represent their thinking', the evidence I will be observing is 'children identify the uses of technology in everyday life and use real or imaginary technologies as props in their play'—although the experience is mainly mathematically focused, the children will be offered a cash register as a prop within their play, and they will also have calculators and pen and paper within their experience.

Requirements

The setting for the experience is a dramatic play. The 'home corner' area will be changed into a hardware store, which is the current interest of the children I am observing. As the children enter the store they will be given an amount of money, which they will be able to use to purchase items within the store. There will be shelves set out in aisles with hardware tools such as hammers, nails, drills, and different tools, enabling the children to browse the various items on offer. The hardware store will have a cash register at the front, which is where the children will purchase their items from a cashier who will have a calculator, cash register, paper, and pens at their register to be able to write out receipts for their customers, and handle and calculate money. Depending on their choice, they may be able to use pen and paper and manually work out the amount of money they are owed and the amount of change required. The hardware store will also have baskets/bags into which the children may place their items, and each item will have a price tag on it, giving the children opportunity to work out how much everything they wish to purchase costs and figure out if they have enough money to purchase what they would like. If they do not have enough money, they will put items back—this will give them an understanding that they cannot spend more than their allocated amount of money. I will need shelving units, a cash register at a bench, bags/baskets, paper money and coins, tools such as hammers, nails, boards of wood, drills, drill bits, a calculator, pens, paper, and catalogues from hardware stores such as Bunnings or Home Mart.

Procedure

The children currently have an interest in money and hardware tools, as many of their family members work within a hardware store and handle money at the cashier's counter. The children have the interest in tools because they spend time with family using different tools to do different things around their house. I will gain the children's interest by exploring the hardware store area with them and showing them what is available to them. I will also ask them different questions about the store, which they have walked through, and I will ask them if they would be interested in looking at the money that they would be using to purchase their tools and resources. After we have looked at the money and the children seem to understand what each note or coin represents, I will offer them the opportunity to explore, and engage with, the hardware store. I will ask them to take turns being the cashier and the customers. I will use strategies for cooperation, such as a timer to give each child a fair and equal turn at each identity. I will explain to the children what the cashier does and what the paper, pens, and calculator are for, using language that they understand. I will observe from afar until a child invites me in; if the cashier is struggling with working out the money, I will ask them if they need help and will give the child small prompts on how to work out their calculation.

If I need to end the experience, I will ask the children to freeze as we do during transition times and to pack away the hardware store—restoring it to the condition in which they found it when first introduced to it. If the children need redirecting for a reason without packing away, I would wait until there is a quieter time for the children and ask a few of them at a time to come with me and pack away the hardware store. If the children were unable to help me pack away the hardware store I would wait until they had all gone home or outside for the day and tidy the area, returning it to its original state in readiness for the next play experience. The area would be set up for all children; however, I would complete my experience during our small group times.

Plan for review

I will determine whether the learning focus for the experience was appropriate by reference to the knowledge that I find that the children have learnt or have not learnt, and I will ask them questions about the value of money and how they look at money, and about the technologies that they used within their experience. I will determine whether the strategies that I used were effective in relation to my learning focus by considering how the experience was handled and explored by the children involved. I will determine that learning and/or development was facilitated if I can observe that the children involved have learnt something new or were challenged during this experience or as a result of their reflection on the experience after they completed it. I would be able to determine if the experience was suitable and successful from observing the children during this experience, seeing whether or not they were able to use a calculator or manually work out the calculations, and seeing how long they were engaged in the experience. I would determine if there were any ways I could improve the experience to make it more successful or appropriate by observing the children and noting what they found interesting and what they did not find interesting. I would also ask the children themselves what they think I could change to make the experience more successful. To determine if my role was appropriate, I would observe whether the children were able to understand what I was explaining to them and if they were interested in what I was saying.

I will be watching for the children's interest and ideas, their speech and comments about the experience as well as their body language, abilities and knowledge, and their development—these elements will help me evaluate the teaching and learning that informed, and resulted from, this experience.

(Source: Sussann Beer)

Chapter summary

This chapter has demonstrated the relationship between assessment and planning in early childhood mathematics education. Authentic assessment provides a foundation upon which to plan meaningful and developmental mathematical learning experiences for children. Children can be actively involved in both planning and assessment processes. Such involvement helps to promote children's engagement in the methods through which they are assessed and in the mathematical learning experiences that are provided for them.

The specific topics we covered in this chapter were:

- assessment and planning
- ELPSARA framework
- documenting and communicating mathematical learning
- photographs
- conversations
- documentation panels
- drawings
- learning stories
- observing and interpreting mathematical learning
- planning for mathematical learning
- Learning Experience Plans.

FOR FURTHER REFLECTION

Assessment in early childhood education settings should be authentic and purposeful. Before assessing children's mathematical learning, consider the following:

1. Why should we assess?
2. What should we assess?
3. For whom should we assess?
4. How should we assess?
5. When should we assess?
6. Who should undertake the assessment?
7. How will the assessment inform our planning?

FURTHER READING

Carr, M. (2001). *Assessment in early childhood settings: Learning stories*. London: Paul Chapman.

Copley, J.V. (1999). Assessing the mathematical understanding of the young child. In J.V. Copley (ed.), *Mathematics in the early years* (pp. 182–8). Reston, VA: National Council of Teachers of Mathematics.

Department of Education, Employment and Workplace Relations. (2010). *Educators belonging, being and becoming: Educators' guide to the Early Years Learning Framework for Australia*. Barton, ACT: Commonwealth of Australia.

MacDonald, A. (2015). Assessment and planning. In *Investigating mathematics, science and technology in early childhood* (pp. 31–51). South Melbourne, VIC: Oxford University Press.

Part 3

CONTENT FOR TEACHING MATHEMATICS IN EARLY CHILDHOOD EDUCATION

OVERVIEW OF PART 3

Part 3 of this book consists of five chapters that examine the mathematical content areas that are developed in early childhood education. Individually, the chapters address the following topics:

» patterns

» space and geometry

» measurement

» number and algebra

» data, statistics, and probability.

Collectively, these chapters focus on explanations of mathematical concepts and show how these concepts can be explored with young children. Part 3 provides a mathematical content knowledge base for your role as early childhood mathematics educator. The information in these chapters will assist you to plan for content-rich mathematical learning experiences and respond appropriately to mathematical learning opportunities as they arise.

12 Patterns

Learn more
Go online to see a video from Paige Lee explaining the key messages of this chapter.

CHAPTER OVERVIEW

This chapter will explore the foundational patterns concepts that are developed in the early childhood years, present examples of these concepts, and provide educator reflections that show how patterns can be explored with young children.

In this chapter, you will learn about the following topics:

» pattern and structure
» recognising patterns
» copying patterns
» continuing patterns
» creating patterns
» repeating patterns
» growing patterns
» symmetrical patterns
» arrays.

KEY TERMS

» Arrays
» Decompose
» Growing patterns
» Pattern
» Repeating patterns
» Structure
» Symmetrical patterns

Introduction

Understanding pattern is essential for the development of mathematical concepts (Knaus, 2013). For example, algebraic reasoning is underpinned by knowledge of pattern and structure, while patterns are evident in the way we use repeated units to measure or represent data. Indeed, mathematics is often referred to as the study of patterns (Knaus, 2013).

'Pattern' is a commonly used term, though it is not always used in a strictly mathematical sense. While children will often produce artworks and proclaim, 'I made a pretty pattern!', it is often the case that these arrangements of colour and shape do not constitute a 'pattern' in the mathematical sense of the word. From a mathematical perspective, a **pattern** can be defined as a sequence of two or more items that begin to repeat themselves. This element of repetition is key to a sequence being recognised as a pattern.

Pattern
can be defined as a sequence of two or more items that repeat.

Research has shown that understandings of pattern develop from very early on (for example, Sarama & Clements, 2009). Indeed, even very young babies recognise patterns in familiar songs, or in habitual movements such as the door opening and their parent entering (Montague-Smith & Price, 2012). Activities such as sorting and classifying objects often foster the beginnings of pattern investigation (Miller, 2019), and pattern ideas can be reinforced through kinaesthetic movement, concrete manipulatives, pictorial representations, and numbers (Warren, Miller & Cooper, 2012). By the age of four to five years, children have more complex understandings of relationships and pattern structures, and they understand how different elements relate to one another. In recognising patterns, children show understanding both the similarities and differences between objects and events, and of the order in which objects and events are sequenced—what came before, what comes after (Montague-Smith & Price, 2012). But these things alone do not constitute familiarity with *pattern*; the key element that must be understood is that of *repetition*.

Patterns occur in everyday life, in both natural and human-made forms, and children are naturally exposed to the many patterns around them. It is because of this that working with patterns can help children to begin to make sense of many different mathematical ideas that they encounter in their worlds (Copley, 2001). Patterns help children to discern structure, identify relationships, and make predictions about what might happen next (Stelzer, 2005). Importantly, patterning skills in early mathematics has been shown to positively impact children's later mathematical achievement (Papic, 2007).

Educator reflection

In my work with children, I am often amused at how they arrange their play items in patterns, whether it is blocks in a tower or dolls in a doll house. When I watch children organise items into patterns or engage them in a nursery rhyme, it is very clear that a certain stimulus is being triggered. Children play with patterns as a way to make sense of mathematical ideas. These ideas help children develop an understanding of mathematical concepts. As an educator, I believe that by having a better understanding as to how interaction with patterns helps develop mathematical awareness in children, we will be able to ensure that programming around patterns is not just limited to an engaging experience but rather provides an opportunity to build an appreciation for concepts around maths.

(Source: Rebecca O'Gorman)

Learn more
Go online to hear this Educator reflection being read.

Opinion Piece 12.1

Learn more
Go online to hear this Opinion Piece being read.

ASSOCIATE PROFESSOR JODIE MILLER

Mathematics is often described as the science of patterns, and it is clear why this may be the case. Patterns form part of our everyday lives and as such we observe and interact with patterns naturally. These interactions help us to explain or predict the world around us. For example, repeating patterns exist in tidal movements, seasonal changes, and geometric patterns. I believe early mathematics starts with patterning contexts as these provide a close home to learning setting connection. In addition, developing capability in patterning is fundamental for early understanding of mathematical structures and concepts.

We know that very young children are capable of engaging in a range of pattern contexts including repeating patterns and growing patterns. These initial learning experiences may begin with sorting and classifying activities. Again there is an opportunity to link experiences young children have in the home to the early learning setting, for example, sorting socks or sorting food types. These experiences give children the opportunity to identify and describe similarities and differences in objects. Following on from this, young children then progress to activities involving repeating and growing patterns, with learning experiences that involve kinaesthetic movement, space, concrete manipulatives, pictures, and numbers.

As educators we need to give experiences to young children that provide opportunities for them to search, notice the uniformity, describe, create, and justify patterns. It is through these early learnings that young children begin to recognise patterns and structures of mathematics which has been found to have a positive influence on a young child's overall achievement in mathematics in later years of schooling.

Jodie Miller is an Associate Professor in mathematics education in the School of Education at the University of Queensland. Her research focuses on improving the educational outcomes for students most at risk of marginalisation in school. Jodie is internationally recognised for her research in early algebraic thinking and evidenced-based strategies to support engagement in mathematics in the early years and primary school settings. She leads research projects with a focus on developing early algebraic reasoning, classroom and mathematical practices, teacher professional development, culturally responsive teaching, and examining student understanding. Prior to working in academia, Jodie was a primary school teacher.

Pattern and structure

Recognising that a sequence of objects makes a pattern, explaining why, and being able to copy, extend, and create a new pattern are all key steps in developing pattern concepts (Montague-Smith & Price, 2012). Understandings of pattern in turn assist children to discern **structure**, or the features and characteristics of the mathematics with which they engage. An ability to discern structure is important for other areas of mathematics, as it assists children with computation and forms the basis of algebraic understanding (Siemon et al., 2011). This section looks at the skills of recognising, copying, continuing, and creating patterns, and explores how they contribute to children's developing awareness of pattern and structure.

Structure refers to the features and characteristics of mathematics.

RECOGNISING PATTERNS

Awareness of different forms of patterns—visual, audible, routines, etc.—is the first step in developing knowledge about different pattern structures. When educators work with children on pattern recognition, the letters of the alphabet are often used to identify the items that make up a pattern. For example, a pattern in its most simple form might consist of two items that repeat. These two items are often termed 'A' and 'B'. As such, a pattern in the form of that depicted in Figure 12.1 is often referred to as an 'A–B pattern'.

FIGURE 12.1 A–B pattern

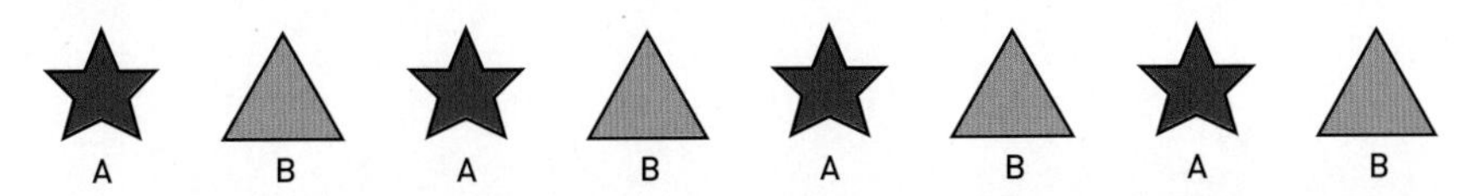

By this logic, it is not too much of a stretch to see how this naming system might be applied to other more complex two-item pattern structures, for example an A–B–B–A pattern (Figure 12.2), or structures involving three or more items, for example an A–B–C pattern (Figure 12.3).

FIGURE 12.2 A–B–B–A pattern

FIGURE 12.3 A–B–C pattern

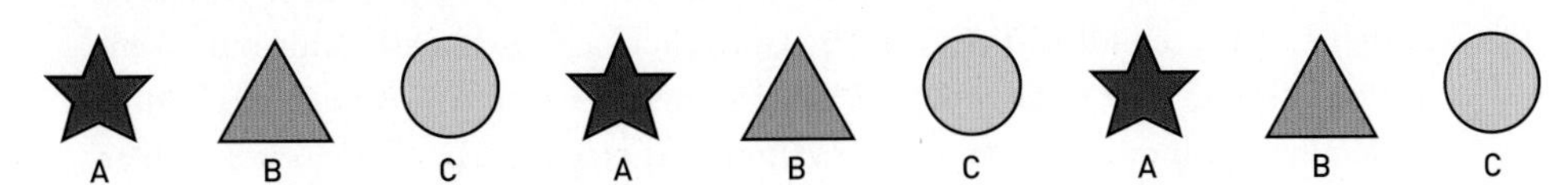

If we have an understanding of the underlying structures of various patterns, it becomes quite easy to recognise and appreciate the patterns that are all around us in our everyday lives. Consider, for example, the many patterns that exist in nature—as Figure 12.4 shows, we can recognise a range of repeating, growing, and symmetrical patterns in natural forms.

FIGURE 12.4 Repeating, growing, and symmetrical patterns in nature

Learn more
Go online to hear this Educator reflection being read.

Educator reflection

Patterns can be found in almost anything and this can be easy to demonstrate to children in early childhood. Quite often preschool children will point out to me patterns in things that I may not have noticed myself, which shows that the children are developing inquisitive and observational characteristics. Sometimes I plan activities to demonstrate patterns using items from the 'maths' box, but I get more satisfaction out of spontaneous activities that are child led, as I enjoy seeing children's imaginations at work.

(Source: Gabrielle Pritchard)

Pause and reflect

Take a look around you. What patterns can you identify? Can you articulate the structure of these patterns?

COPYING PATTERNS

Copying patterns basically requires an ability to **decompose**—that is, a pattern must be reduced to its individual parts, and each part must be matched on a one-for-one basis. This initially requires the visual discrimination of the individual components that make up the pattern. In performing this discrimination, children will check back to the original pattern, using visual skills to compare the elements of their copied pattern. Though not strictly patterning, activities such as building Lego using an instruction booklet (Figure 12.5) are a great way for children to practise discerning and matching the characteristics of objects, which is a skill essential to copying patterns.

Decompose means to break a pattern down to its individual parts.

FIGURE 12.5 Matching the Lego blocks to the instructions

CONTINUING PATTERNS

The basic premise of understanding continuing patterns is that the observer *knows what comes next*. However, in order to know what comes next, children must first be able to discern the repeating structure of the pattern that has been given thus far and to identify which repeating elements must be utilised to continue the pattern. It may be helpful to encourage children to pause each time they make a repeat and identify the repeating element, before continuing to the next repeat (Siemon et al., 2011), thus reinforcing the notion of the repeating component. The ability to continue patterns assists the development of additive, multiplicative, and algebraic thinking skills.

CREATING PATTERNS

When exploring the creation of patterns with young children, it is important to remember that patterns have many different attributes and can be constructed from a vast range of objects and events. While concrete materials such as coloured blocks and beads (or indeed fruit, as shown in Figure 12.6) can be useful resources for developing understandings of patterns, it is important that we take care not to limit patterning experiences to these sorts of activities. By understanding the many attributes associated with patterning—colour, shape, size, texture, position, quantity—we can appreciate the diversity of patterns that might be constructed.

FIGURE 12.6 Creating a simple repeating pattern

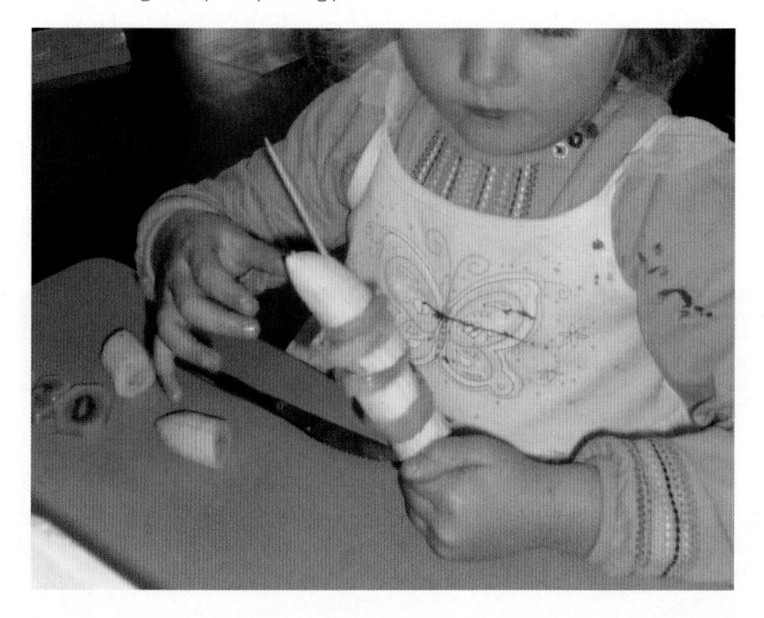

(Source: Michelle Muller)

CURRICULUM CONNECTIONS

Early Years Learning Framework

Outcome 5: Children are effective communicators

Children begin to understand how symbols and pattern systems work.

This is evident when children:

- Begin to make connections between and see patterns in their feelings, ideas, words and actions and those of others
- Notice and predict the patterns of regular routines and the passing of time
- Begin to recognise patterns and relationships and the connections between them
- Listen and respond to sounds and patterns in speech, stories and rhyme
- Draw on memory of a sequence to complete a task.

Educators promote this learning when they:

- Draw children's attention to symbols and patterns in their environment and talk about patterns and relationships, including the relationship between letters and sounds
- Provide children with access to a wide range of everyday materials that they can use to create patterns and to sort, categorise, order and compare.

(DET, 2019, p. 46)

CURRICULUM CONNECTIONS

Australian Curriculum: Mathematics

Foundation year:

Copy, continue and create patterns with objects and drawings (ACMNA005)

- Observing natural patterns in the world around us
- Creating and describing patterns using materials, sounds, movements or drawings.

(ACARA, 2022)

Types of patterns

As outlined in the Introduction to this chapter, the first step in understanding patterns is knowing that they are constituted by sequences of items that begin to repeat themselves. This element of repetition is key; without it, we have an arrangement of items, but not a pattern.

Patterns may be constructed in a number of ways, but they generally take one of three forms: 1. repeating pattern; 2. growing pattern; or 3. symmetrical pattern. A further pattern structure is the array, which may contain any combination of these three forms. Each of these forms will now be explored in greater depth.

REPEATING PATTERNS

As the name suggests, **repeating patterns** are patterns made up of repeated sequences of items (which include objects, shapes, colours, sounds, events, and so forth). These patterns may be constructed in different layouts, such as linear (Figure 12.7) or circular (a rearrangement of a linear form) (Figure 12.8); alternatively, they may be laid out in repeating rows or columns (Figure 12.9), or in diagonal arrangements (Figure 12.10).

Repeating patterns are repeated sequences of items, which may be constructed using different layouts.

FIGURE 12.7 Linear repeating pattern

FIGURE 12.8 Circular repeating pattern

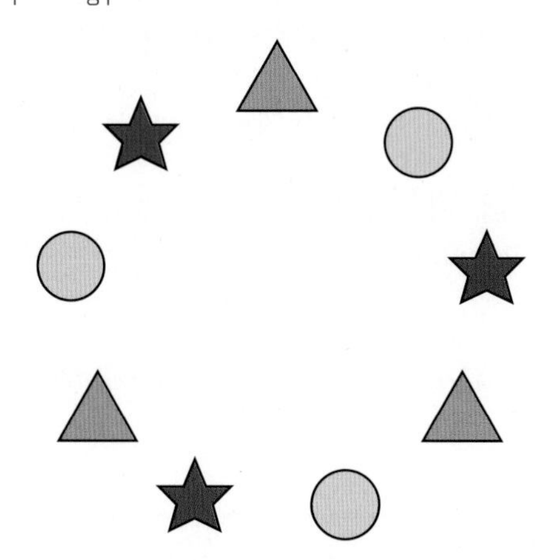

FIGURE 12.9 Repeating patterns displayed in rows (i) and columns (ii)

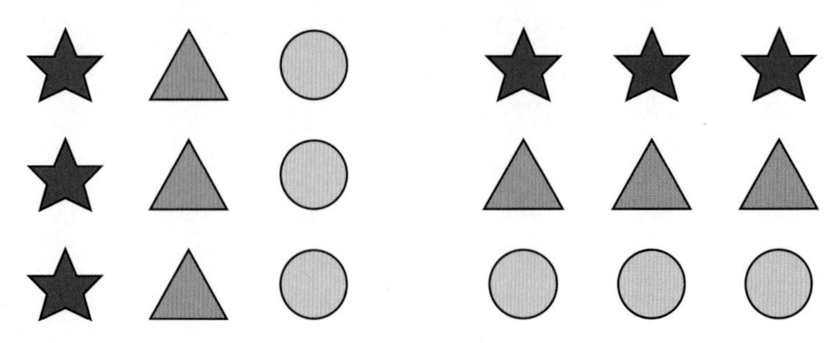

FIGURE 12.10 Diagonal repeating patterns

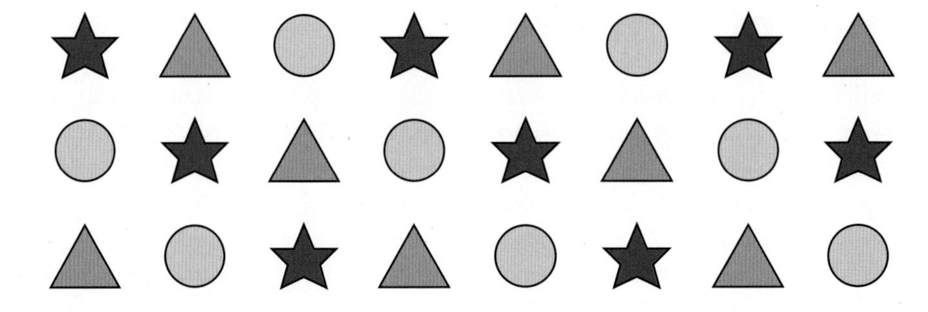

GROWING PATTERNS

Growing patterns are those that have a similar relationship between elements, but in which each element increases or decreases.

Growing patterns are those in which a similar relationship is maintained between one element and the next, but in which the shape or number increases or decreases in size (Montague-Smith & Price, 2012). For example, the number sequence 1, 3, 5, 7 ... is an example of a growing pattern because each element of the pattern increases by two each time.

CURRICULUM CONNECTIONS

Australian Curriculum: Mathematics

Year one:

Investigate and describe number patterns formed by skip-counting and patterns with objects (ACMNA018)

- Using place-value patterns beyond the teens to generalise the number sequence and predict the next number
- Investigating patterns in the number system, such as the occurrence of a particular digit in the numbers to 100.

Year two:

Describe patterns with numbers and identify missing elements (ACMNA035)

- Investigating features of number patterns resulting from adding twos, fives or tens.

(ACARA, 2022)

SYMMETRICAL PATTERNS

Symmetrical patterns are formed on the basis of the reflective or rotational correspondence of the item(s) that make up the pattern. Symmetry will be addressed in greater detail in Chapter 13, but for the purpose of understanding its relationship to patterning, it is helpful to recognise that patterns often contain items that 'mirror' one another in various ways. Consider Figure 12.11 and Figure 12.12, which, respectively, display real-world examples of patterns based on reflective symmetry and on rotational symmetry.

Symmetrical patterns are formed on the basis of the reflective or rotational correspondence of items.

FIGURE 12.11 Reflectively symmetrical patterns

FIGURE 12.12 Rotationally symmetrical patterns

CURRICULUM CONNECTIONS

Australian Curriculum: Mathematics

Year two:

Identify and describe half and quarter turns (ACMMG046)

- Predicting and reproducing a pattern based around half and quarter turns of a shape and sketching the next element in the pattern.

(ACARA, 2022)

ARRAYS

Arrays
are formed as an arrangement of rows and columns with equal numbers in each row and equal numbers in each column.

Patterns are also evident in **arrays**, which are structures configured by an arrangement of rows and columns with equal numbers in each row and equal numbers in each column (see Figure 12.13). It is particularly important that children become familiar with arrayed structures, as these patterns will assist in later learning of multiplication (see Chapter 15) and of the calculation of areas (see Chapter 14).

FIGURE 12.13 An example of an array

Pause and reflect

What other examples of arrays in everyday life can you think of?

Chapter summary

In this chapter I have presented the pattern concepts that can be explored in early childhood mathematics education. Patterning concepts and skills are critical for other domains of mathematical learning, such as space, measurement, and number. It is essential that children are given a variety of opportunities to develop knowledge of patterning through exploration of everyday objects and environments.

The specific topics we covered in this chapter were:

- pattern and structure
- recognising patterns
- copying patterns
- continuing patterns
- creating patterns
- repeating patterns
- growing patterns
- symmetrical patterns
- arrays.

FOR FURTHER REFLECTION

Examine Figure 12.14, which shows a kindergarten classroom. Consider the following:

1 What patterns might be noticed in this classroom?
2 How might children be prompted to notice these patterns?
3 What resources in this image might be used to recognise, copy, continue, and create patterns?

FIGURE 12.14 A kindergarten classroom

(Source: Amy MacDonald)

FURTHER READING

Jorgensen, R., & Dole, S. (2011). Patterns and algebra. In *Teaching mathematics in primary schools* (2nd ed., pp. 258–74). Crows Nest, NSW: Allen & Unwin.

Knaus, M. (2013). Geometry—spatial awareness and shape. In *Maths is all around you: Developing mathematical concepts in the early years* (pp. 61–76). Albert Park, VIC: Teaching Solutions.

Knaus, M. (2013). Pattern. In *Maths is all around you: Developing mathematical concepts in the early years* (pp. 22–32). Albert Park, VIC: Teaching Solutions.

Montague-Smith, A., & Price, A.J. (2012). Pattern. In *Mathematics in early years education* (3rd ed., pp. 83–114). New York: Routledge.

Montague-Smith, A., & Price, A.J. (2012). Shape and space. In *Mathematics in early years education* (3rd ed., pp. 115–44). New York: Routledge.

13 Space and Geometry

Learn more
Go online to see a video from Paige Lee explaining the key messages of this chapter.

CHAPTER OVERVIEW

This chapter will explore the foundational space and geometry concepts that are developed in the early childhood years, present examples of these concepts, and provide example Learning Experience Plans that show how space and geometry can be explored with young children.

In this chapter, you will learn about the following topics:

- spatial reasoning
- spatial orientation
- spatial visualisation
- van Hiele levels of geometric thinking
- shapes
- 2D and 3D shapes
- attributes of shapes
- lines
- symmetry
- transformations
- tessellation
- position and direction
- mapping.

KEY TERMS

- 2D shapes
- 3D shapes
- Congruent transformations
- Edge
- Face
- Geometric reasoning
- Geometry
- Line of symmetry
- Polygons
- Projective transformations
- Rotational symmetry
- Space
- Spatial orientation
- Spatial reasoning
- Spatial visualisation
- Symmetry
- Tessellation
- Topological transformations
- Transformation

Introduction

This chapter outlines the big ideas associated with space and geometry. 'Space' is the area of mathematics that develops into 'geometry' in the later years of school mathematics (Montague-Smith & Price, 2012). **Space** refers to an understanding of the properties of objects, as well as the relationships between objects. Similarly, the term **geometry** is used to describe the domain of mathematics that encompasses size, shape, relative position, and movement of two-dimensional figures in the plane and three-dimensional objects in space (ACARA, 2022). In short, space and geometry are about understanding shapes and figures and their relationships, including the ways in which shapes and figures can be transformed, and the ways in which they move through and around spaces (Knaus, 2013).

Space
refers to an understanding of the properties of objects, as well as the relationships between objects.

Geometry
encompasses size, shape, relative position, and movement of two-dimensional figures in the plane and three-dimensional objects in space.

Spatial awareness and geometric reasoning are necessary for existing in environments, as these understandings allow us to process and interpret the spatial cues and structures that are all around us, in both natural and built environments. Spatial knowledge helps us both to understand, and to communicate about, our worlds. Through the use of representations such as drawings, photographs, maps, and plans, we can document, communicate, and interpret spatial relationships.

Research has shown that understandings of space are innate to young children (for example, Sarama & Clements, 2009). Babies begin to recognise shapes and navigate spaces, and by two to three years of age, children recognise and use symmetry, identify a variety of shapes, and learn about positional concepts and relationships (Montague-Smith & Price, 2012). By four to five years, children recognise and represent a wide range of two-dimensional (2D) shapes, and they begin to understand 2D shapes as being the faces of three-dimensional (3D) shapes. By this stage, children have attained more complex understandings of spatial relationships and structures, and they understand how different elements relate to one another.

The ability to develop and use spatial skills is dependent upon the development of understandings about discrete aspects of space and geometry, and how they relate to one another. While there are many facets of spatial understanding, this chapter explores those that develop most readily in the early childhood years. The chapter begins with an overview of spatial reasoning, before exploring concepts associated with shapes, lines, symmetry, and position, direction and orientation. This chapter also discusses spatial relationships and looks at how these relationships might be explored through transformations and mapping.

Opinion Piece 13.1

ASSOCIATE PROFESSOR KAY OWENS

When I returned to Australia a couple of decades ago, I wondered why teacher educators were enthusiastic about geometry activities such as making pentominoes and using tangrams. Then I learnt from children by observing them that they were using visuospatial reasoning to solve problems and that these kinds of activities

Learn more
Go online to hear Associate Professor Kay Owens reading this Opinion Piece.

assisted children to develop their important ability to visualise and think about shapes. My further study of how visualising and spatial abilities were important in mathematics indicated that children could learn these skills, and my research in this area has continued. Visualising and spatial abilities are malleable and important for geometry. This importance, indeed, has been underestimated. I found that teachers were uneasy teaching geometry because they just thought it was all about learning the names of all the shapes. Far from it, children should make shapes—2D shapes, different triangles, different four-sided shapes cut from paper, made with string, elastic, or sticks; and they should fashion shapes from other shapes. Children should be encouraged to make a good variety of each shape so that they are not limited by a restricted image—for example, if they are making triangles, their outputs should not be limited to equilateral triangles. When making 3D shapes, let them turn their ball of play dough into an object with four, five, or six flat faces and count the edges where the flat faces meet. They'll use their dynamic imagery to solve the problems that they encounter.

Another area of research that I favour started with Pirie and Kieren's (1994) work that indicated children would move from their initial knowing to make and hold images about which they would notice properties and formalise their concepts. If necessary, they would return to the image making or holding, or notice other properties. This visuospatial reasoning is important in learning concepts. Some research studies are critical of stage theories like the van Hiele stages, which are considered too linear for us to gain a deep understanding of children's learning of geometry. Visuospatial reasoning builds on the emerging imagery—concrete or pictorial imagery—that children possess, which enables them to use pattern and dynamic imagery to develop their concepts and solve problems (Owens, McPhail, & Reddacliff, 2003).

Kay Owens is an Adjunct Associate Professor of Mathematics Education at Charles Sturt University, Dubbo, Australia. Kay worked for fifteen years in Papua New Guinea, where she held a lectureship in mathematics at the PNG University of Technology and was a head of department at Balob Teachers College. Kay has continued her research, working with colleagues in Papua New Guinea and Sweden in such areas as ethnomathematics (encompassing mathematics, language, and culture) and space, geometry, and measurement education.

Spatial reasoning

Spatial reasoning is the ability to see, inspect, and reflect on spatial objects, images, relationships, and transformations.

Children rely on spatial reasoning processes to engage with, understand and apply concepts associated with, geometry (Bobis, Mulligan, & Lowrie, 2013). **Spatial reasoning** can be defined as 'the ability to "see", inspect, and reflect on spatial objects, images, relationships and transformations' (Battista, 2007, p. 843). For example, the ability to recognise the wheels of a car as 'circles' is an application of spatial reasoning. Children's toys such as the classic Tupperware shape sorter ball, and stacking rings (pictured in Figure 13.1), are designed to promote spatial reasoning from a very young age. As these skills become more formalised, children develop **geometric reasoning**, which has been described by Battista (2007, p. 843) as 'the invention and use of formal conceptual systems to investigate shape and space'. For example, in the early primary years, children begin to use formal units and tools to represent shapes and spaces, and they begin to understand the properties of these shapes and spaces.

Geometric reasoning is the invention and use of formal systems to investigate shape and space.

Current research supports the importance of these visuospatial skills, with studies showing that spatial reasoning is one of the most important contributors to success in mathematics and in technical fields reliant upon mathematical knowledge (for example, Lowrie, Logan, & Ramful, 2016).

FIGURE 13.1 Toys that promote spatial reasoning

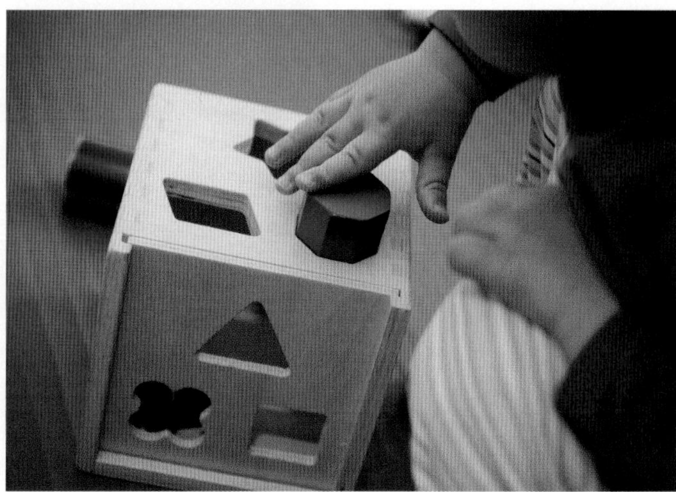

What other toys/play resources can you think of that promote children's spatial reasoning? How can these resources be effectively utilised in an early childhood program?

Pause and reflect

SPATIAL ORIENTATION

Spatial orientation can be defined as knowing where you are and how to get around (Clements, 1999a; Siemon et al., 2021). For example, if you use Google Maps to work out how to get from your house to a friend's house, you are putting your spatial orientation skills into action. Children develop spatial orientation from a very young age (indeed, it might be argued, from birth), as they learn to navigate their space. Consider, for example, the spatial orientation skills required for an infant to navigate a play space such as that pictured in Figure 13.2. Young children learn practical navigation skills early in their lives (Clements, 1999a).

Spatial orientation means knowing where you are and how to get around.

FIGURE 13.2 Navigating space

(Source: Amy MacDonald)

Educator reflection

Learn more
Go online to hear this Educator reflection being read.

I have spent a lot of time observing the way children move themselves around the yard and how aware they are of their surroundings as they play. It amazes me to see the children carrying large objects and manoeuvring them around their peers in order, for example, to take the log where they want to go, or to see the children pedalling their little hearts out as they race around on their bikes and watch carefully where their friends are so they don't run into them.

(Source: Sarah Morrow)

CURRICULUM CONNECTIONS

Early Years Learning Framework

Outcome 3: Children have a strong sense of wellbeing

Children take increasing responsibility for their own health and physical wellbeing

This is evident when children:

- Use their sensory capabilities and dispositions with increasing integration, skill and purpose to explore and respond to their world
- Demonstrate spatial awareness and orient themselves, moving around and through their environments confidently and safely.

Educators promote this learning when they:

- Plan for and participate in energetic physical activity with children, including dance, drama, movement and games
- Provide a wide range of tools and materials to resource children's fine and gross motor skills.

Outcome 4: Children are confident and involved learners

Children develop dispositions for learning such as curiosity, cooperation, confidence, creativity, commitment, enthusiasm, persistence, imagination and reflexivity.

This is evident when children:

- Express wonder and interest in their environments
- Participate in a variety of rich and meaningful inquiry-based experiences.

Educators promote this learning when they:

- Provide learning environments that are flexible and open-ended.

(DET, 2019, pp. 35, 37)

SPATIAL VISUALISATION

Visualisation is an important aspect of geometry and of mathematics in general (Siemon et al., 2021). Information is frequently presented visually, particularly in the form of diagrams and graphics (Bobis et al., 2013). Indeed, children are exposed to a variety of visual stimuli in their daily lives, and they need to develop visual skills to make sense of the world around them (Bobis et al., 2013).

Spatial visualisation is the ability to generate and manipulate a mental image or representation of spatial relationships. To build on the earlier example of using Google Maps, you might—when driving from your house to your friend's house—retain a mental picture of the map you viewed in order to help you navigate this journey. Young children build mental representations to assist them in navigating spaces, such as remembering the location of a favourite toy, or knowing in which direction to crawl to avoid bumping into the table leg.

Spatial visualisation is the ability to generate and manipulate a mental image or representation.

Gutiérrez (1996, cited in Siemon et al., 2011, pp. 486–7) proposed a framework of six main abilities that constitute spatial visualisation:

1. *figure-ground perception*—the ability to identify and isolate a specific figure out of a complex background
2. *perceptual constancy*—the ability to recognise that some characteristics of an object are independent of size, colour, texture, or position
3. *mental rotation*—the ability to produce dynamic mental images and visualise movement of a mental image
4. *perception of positions*—the ability to relate objects, pictures, or mental images to oneself
5. *perception of spatial relationships*—the ability to relate several objects to each other and to the person concerned
6. *visual discrimination*—the ability to compare several figures and determine how they are similar or different.

Pause and reflect

Consider your own spatial visualisation skills: Would you describe this as an area of strength, or an area for development? What everyday opportunities for developing your spatial visualisation skills can you think of?

THE VAN HIELE LEVELS OF GEOMETRIC THINKING

Understanding space and geometry generally entails understanding properties of objects, relationships between objects, and the position, location, and orientation of objects within space. In the 1980s, researchers Pierre and Dina van Hiele proposed that children developed their geometric thinking by moving through discrete stages of reasoning (Siemon et al., 2011). These stages, which came to be known as the 'van Hiele levels', are displayed in Table 13.1.

TABLE 13.1 van Hiele levels of geometric thought

LEVEL 0	*Pre-recognition* » Children attend only to some of a shape's characteristics. » Children are unable to identify many common shapes.
LEVEL 1	*Visual* » Children identify and operate on geometric shapes according to their appearance. » Children recognise shapes as whole structures and rely heavily on memorised examples.
LEVEL 2	*Descriptive/Analytic* » Students recognise and can classify shapes by their properties.
LEVEL 3	*Abstract/Relational* » Students can form definitions, distinguish between necessary and sufficient sets of conditions, and understand logical arguments. » Some students may be able to construct logical arguments.
LEVEL 4	*Formal deduction* » Students establish theorems.
LEVEL 5	*Rigour/Metamathematical* » Students reason formally about mathematical systems.

(Source: Siemon et al., 2011)

While the van Hiele model is useful for identifying different elements of children's geometric thinking, the model has been heavily critiqued for the lock-step manner in which it presents children's geometric development. As highlighted by Kay Owens in Opinion Piece 13.1, this linear model of geometric development limits our capacity to understand the fluid ways in which children acquire spatial and geometric knowledge. As such, rather than consider space and geometry from the perspective of linear developmental levels, it is perhaps more useful for us to think about how children develop knowledge of space and geometry in a holistic and dynamic manner.

Pause and reflect

What do you see as the strengths and weaknesses of the van Hiele levels? How might this model be used effectively in early childhood mathematics education?

Shapes

Informal learning about shapes occurs in the everyday experiences of young children as they start to recognise basic shapes in their environment and become familiar with the words used to label those shapes (Knaus, 2013). Children need physical interaction with objects to be able to connect their actions and create ideas about shapes (Siemon et al., 2011) (Figure 13.3). Clements (1999, p. 67) reminds us that

> Children's ideas about shapes do not come from passive looking. Instead, they come as children's bodies, hands, eyes ... and minds ... engage in action ... Children need to explore shapes extensively to understand them fully. Merely seeing and naming pictures is insufficient ... They have to explore the parts and attributes of shapes.

In short, children should be provided with opportunities to develop practical knowledge about shapes and their relationships; initially through examining shape and structure in the children's environment, and then more formally through formulating relationships between the various properties of shapes (Bobis et al., 2013).

FIGURE 13.3 Exploring shapes through everyday play

(Source: Paige Lee)

Opinion Piece 13.2

ASSOCIATE PROFESSOR KAY OWENS

We need to give children plenty of time to use their play and inquisitiveness to investigate shapes, their properties, and their relationships. For example, children might overlay shapes to compare and discuss angles (this usually occurs around age seven, as children may not be able to distinguish the angles or sides of a shape

Learn more
Go online to hear Associate Professor Kay Owens reading this Opinion Piece.

from the rest of the shape before then—that is the spatial ability of re-seeing or disembedding). Sorting shapes into subgroups is fun and children learn to describe the various critical features of shapes that help to define each category. More interesting is exploration of the symmetries and diagonals of 2D shapes. I like to begin with two diagonals tied in the middle to make different rectangular house plans (an idea from an African group). Names and classifications are culturally determined so we could have quite different shape names and approaches to geometry depending on our cultural backgrounds, and it's even possible for us to change the definitions of geometrical terms. For geometry, our schools tend to follow the approach of the ancient Greeks, which spread through Europe during the Renaissance—it's an approach that happens to be useful for building our houses and schools. However, transformation geometry will encourage dynamic imagery for concept development, and there are plenty of online programs to assist you in introducing this approach to children. So, it is easy to avoid the drill and practice on labels for shapes, and instead to make geometry much more fun by using real and online learning objects.

2D AND 3D SHAPES

2D shapes
are flat shapes that have length and width but no height.

3D shapes
are solid shapes that have length, width, and height.

In the early childhood years, children are exposed to a range of 2D and 3D shapes through their everyday experiences. **2D shapes** are flat shapes that have length and width but no thickness (height)—for example, triangles, squares, circles. **3D shapes** are solid shapes that have length, width, and height—for example, cubes, spheres, cylinders. There is a close relationship between 2D and 3D shapes, because 3D shapes are in fact made up of 2D shapes. Consider the cylinder: it consists of two circles (at either end), which are joined by a rectangle that wraps around the width of the shape.

It is often thought that children should learn about 2D shapes first; however, there is logic underpinning recent arguments that children should learn about 3D shapes in tandem with 2D or, indeed, even *prior* to learning about 2D. This is because a child's world is made up of 3D objects, so it may make more sense for them first to learn the names of the 3D objects with which they are familiar, and then deconstruct these objects to identify the 2D shapes that make them up. Little research has been undertaken into this matter, so the debate may continue to rage for some time yet—in the meantime, I suggest that you take your lead from the children with whom you work and do what you think is best for them!

What is your view in relation to the teaching of 2D and 3D shapes? Which do you believe should come first, and why? How might you prompt children to explore 2D and 3D shapes together in meaningful ways?

CURRICULUM CONNECTIONS

Australian Curriculum: Mathematics

Foundation year:

Sort, describe and name familiar two-dimensional shapes and three-dimensional objects in the environment (ACMMG009)

- Sorting and describing squares, circles, triangles, rectangles, spheres and cubes.

(ACARA, 2022)

ATTRIBUTES OF SHAPES

In addition to helping children to recognise and name 2D and 3D shapes, it is important for educators to provide more detailed explorations of the attributes of shapes (Clements, 1999a; Knaus, 2013). In the early years, children should be exposed to a variety of 2D shapes and 3D objects so that they can begin to make sense of attributes of shapes and structures (Bobis et al., 2013). Children need to be aware of the specific features of different shapes. For example, when exploring 3D shapes children should learn that the **face** is the side of a solid shape, while the **edge** is where two faces meet (see Figure 13.4).

A **face** is the side of a solid shape.

The **edge** of a shape is where two faces meet.

Polygons are any 2D shape with straight sides.

It is also important for children to learn that the names of shapes are more than just labels—each name signifies the properties of the shape that it denotes. For example, the name *triangle* literally means 'tri-angle' or 'three angles'. Indeed, the naming conventions of all **polygons** (any 2D shape with straight sides) are used to indicate the number of sides of the shape—for example, *pentagon* means 'five sides', *hexagon* means 'six sides', and so forth. Some common polygons are shown in Table 13.2.

FIGURE 13.4 Faces and edges of a 3D shape

TABLE 13.2 Some common polygons

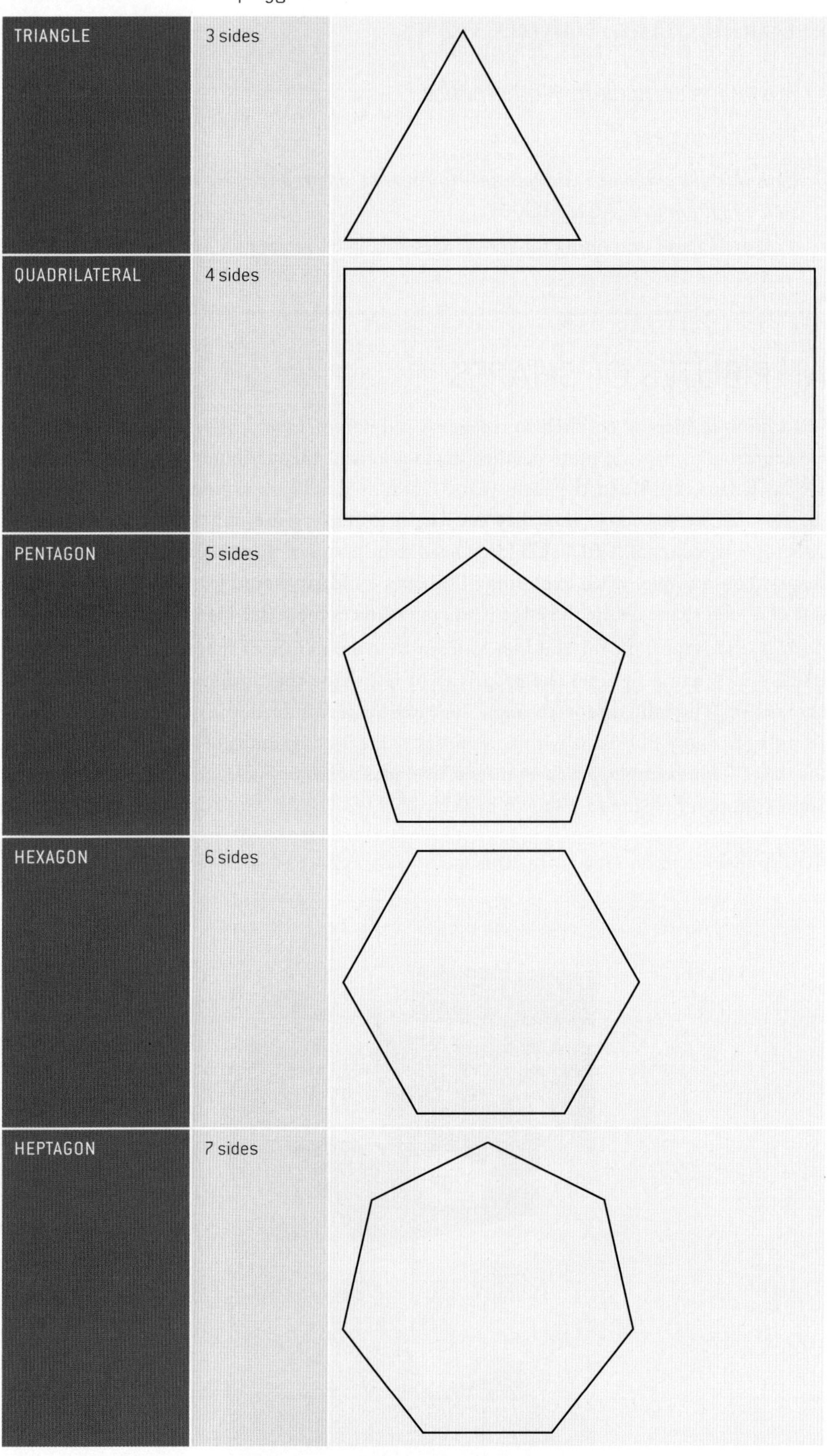

TRIANGLE	3 sides	
QUADRILATERAL	4 sides	
PENTAGON	5 sides	
HEXAGON	6 sides	
HEPTAGON	7 sides	

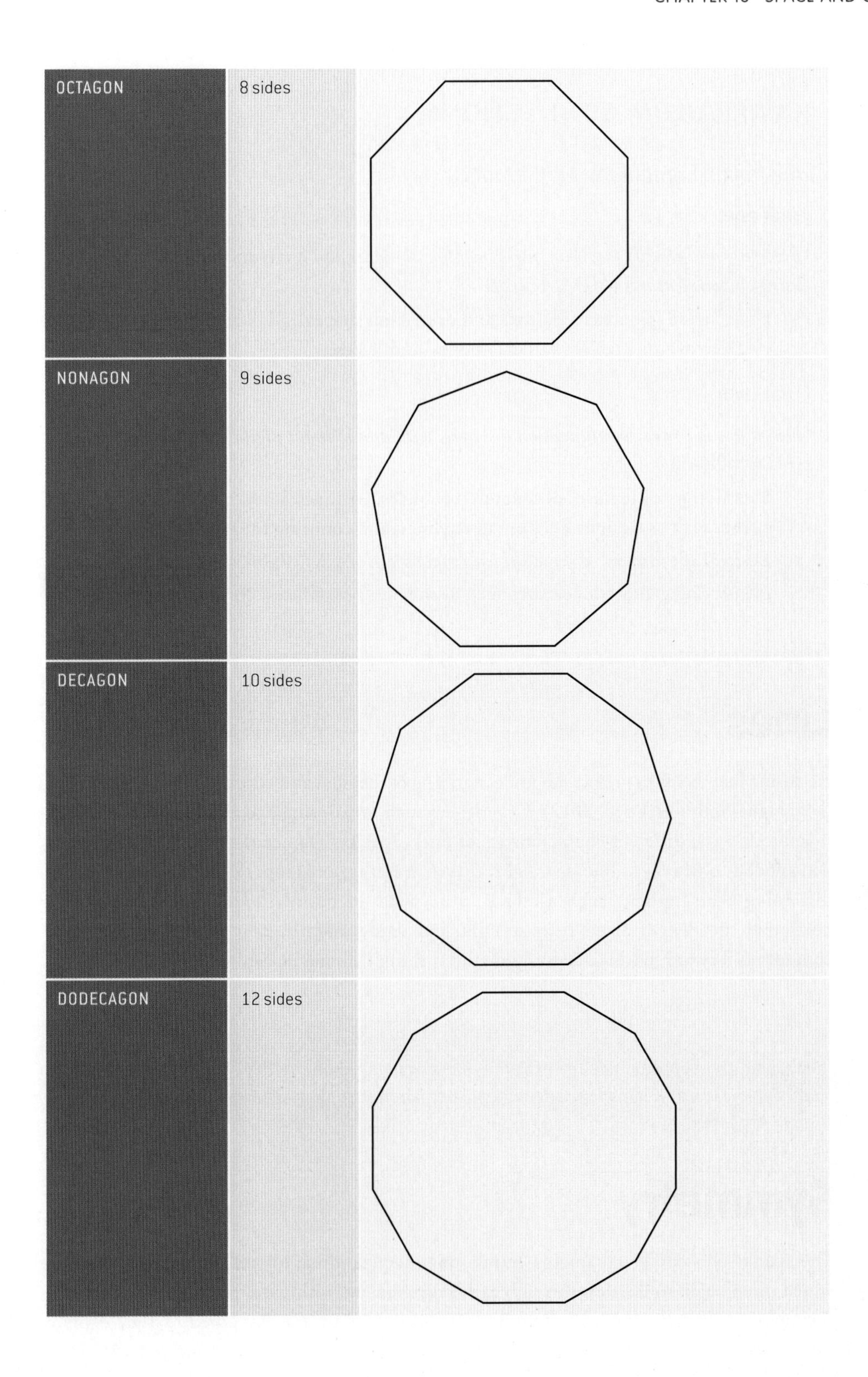
OCTAGON
8 sides
NONAGON
9 sides
DECAGON
10 sides
DODECAGON
12 sides

CURRICULUM CONNECTIONS

Australian Curriculum: Mathematics

Year one:

Recognise and classify familiar two-dimensional shapes and three-dimensional objects using obvious features (ACMMG022)

- Focusing on geometric features and describing shapes and objects using everyday words such as 'corners', 'edges' and 'faces'.

Year two:

Describe and draw two-dimensional shapes, with and without digital technologies (ACMMG042)

- Identifying key features of squares, rectangles, triangles, kites, rhombuses and circles, such as straight lines or curved lines, and counting the edges and corners.
  Describe the features of three-dimensional objects (ACMMG043)
- Identifying geometric features such as the number of faces, corners or edges.

(ACARA, 2022)

Lines

Exploring lines helps to provide an understanding of the sides and edges of 2D and 3D shapes (Montague-Smith & Price, 2012). Experiences such as painting and drawing help children to appreciate the different characteristics of lines. Specific characteristics include *line shape*, which refers to whether a line is straight, curved, a zigzag, or a loop; *thickness*, which involves discerning characteristics such as thick, thin, wide, or narrow; *outlines*, which concerns developing the ideas of shape, turn, straight, and bendy; and *direction*, which addresses such concepts as forward, backward, straight, and turning (Montague-Smith & Price, 2012).

Think about the opportunities for exploring lines while children are drawing. How might you model appropriate mathematical language while children are engaged in drawing activities?

Symmetry

The spatial concept of symmetry builds on understandings of reflection and rotation, which often begin to develop through early engagements with patterning (see Chapter 12). For many children, early understandings about reflection are developed as they explore mirrors—most young children are fascinated with their own reflection (Figure 13.5).

Symmetry
refers to exactly similar parts facing each other or assembled around an axis.

The idea of reflection, at its simplest, is 'something facing itself'. Building on this notion, **symmetry** refers to an object being made up of exactly similar parts facing each other or around an axis. General forms of symmetry include line symmetry, rotational symmetry in 2D space, and symmetry in 3D space.

SYMMETRY IN 2D SPACE

Symmetry in 2D space can arise in two ways—line symmetry and rotational symmetry (Jorgensen & Dole, 2011). The most easily recognisable kind of symmetry, and the best starting point for young children, is the idea of a mirror image. Symmetry of this kind is known as line or lateral symmetry, and is associated with a **line of symmetry**, which refers to a mid-point of a 2D shape at which one side of the shape reflects the other. Some examples are shown in Figure 13.6.

A **line of symmetry** refers to a point of a 2D shape at which one side of the shape reflects the other.

FIGURE 13.5 Exploring reflection

(Source: Amy MacDonald)

FIGURE 13.6 Lines of symmetry

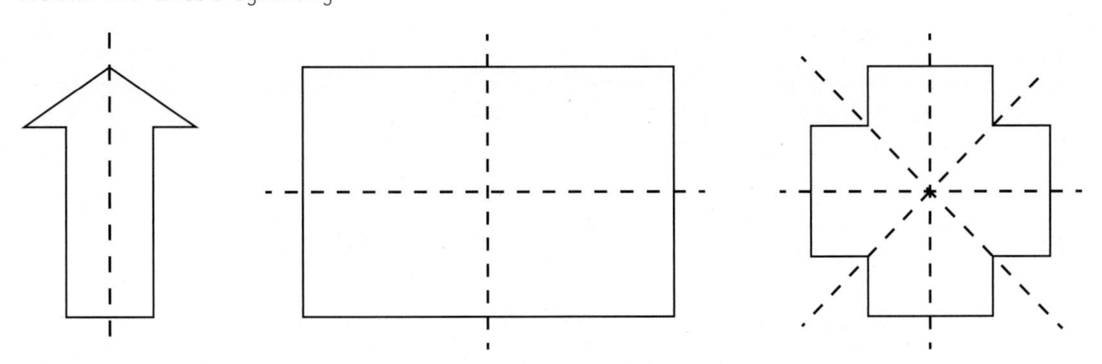

Rotational symmetry is a more challenging concept to grasp and is associated with turning a shape through a fraction of a full circle. For example, a figure is said to have rotational symmetry if, when turned through a fraction of a circle such as one-half (180°), the image corresponds exactly to the original object in terms of the position it occupies (Jorgensen & Dole, 2011). Useful examples of real-world objects with rotational symmetry include things such as pinwheels, propellers, and hubcaps or mags (Figure 13.7).

Rotational symmetry is associated with turning a shape through a fraction of a full circle.

FIGURE 13.7 Rotational symmetry

(Source: Amy MacDonald)

Pause and reflect

What are some other examples of real-world objects that can promote understandings of symmetry? How might you prompt conversations about symmetry during everyday explorations in natural and built environments?

Learning Experience Plan 13.1

'Paper aeroplanes'

Experience information

This experience is planned for four preschool children who have been observed as a group over the last two weeks. The children's ages are as follows:

Z, 4 years 5 months
Q, 4 years 9 months
V, 4 years 2 months
X, 4 years 7 months.

Learning focus

The learning focus for this experience is based on the observations that were taken during the last two weeks. While observing the children, I noticed a common interest between them, which was flying transport. This became the topic for the experience, as I knew the children would be eager to participate. I also chose to use this learning experience for this group not only because of their interest but also because it ties into our latest room project of space and rocket ships, which will make this experience a good follow-on activity.

Mathematical learning

During the construction of paper aeroplanes, there are many concepts that I hope to facilitate, which include learning about shapes and space. As they are folding their paper to make their aeroplanes, the children are learning about symmetry and revisiting their understanding of, and knowledge about, 2D shapes. During this stage, they are also being exposed to the concept of lines and angles, which form part of geometry. When it is time to put their aeroplanes in action, the children will be given the opportunity to interact with the content areas of measurement, and will be calculating and problem solving with numbers. The children will be able to put their knowledge about linear measurement to practise as they measure how far their aeroplanes have travelled by using a tape measure. When they have completed that part of the experience, the children will be encouraged to use the concept of the simplest structure of addition. Each child will have the opportunity to land their aeroplane on a number target, and by doing such landings each child will be encouraged to add up the two score points achieved by their aeroplane on its two flights to the target. They will also have an opportunity to write their scores on the cardboard.

Scientific learning

Although this experience is mainly based on mathematical learning, there are still some scientific concepts to which the children will be exposed. Through group discussions, indirect teaching, and the practical side of this experience, the children will learn and further develop their knowledge about different physical science concepts such as the effects of gravity and air, which will be demonstrated when they fly their planes. The children will also be given the opportunity to extend and share their knowledge about the content area of astronomy, in regard to Newton's three laws of motion.

EYLF outcomes

In 'Learning Outcome 1: Children have a strong sense of identity', the children will be working towards the following areas:

- 'Children feel safe, secure and supported'—this will be evident through the positive interactions the children have with their educators and peers as well as responding to ideas and suggestions from others.
- 'Children develop their emerging autonomy, inter-dependence, resilience and sense of agency', through being open to new challenges and discoveries about how planes work and how to make paper aeroplanes, as well as working collaboratively as a group and taking turns. The children will also be encouraged to persist when faced with challenges and if their first attempts are not successful.

In 'Learning Outcome 3: Children have a strong sense of wellbeing', the children will be working towards the following areas:

- 'Children become strong in their social and emotional wellbeing'—the children will be encouraged and presented with an opportunity to make choices, accept challenges, take considered risks, and cope with frustrations and the unexpected.
- 'Children take increasing responsibility for their own health and physical wellbeing'—the children will be using their spatial awareness and orient themselves as they manipulate their paper aeroplanes with increasing competence and skill in the physical environment.

In 'Learning Outcome 4: Children are confident and involved learners', the children will be working towards the following areas:

- 'Children develop dispositions for learning such as curiosity, cooperation, confidence, creativity, commitment, enthusiasm, persistence, imagination and reflexivity'—the children will be participating in a rich and meaningful inquiry-based experience and will be encouraged to persist even if they find the task difficult.

- 'Children develop a range of skills and processes such as problem solving, inquiry, experimentation, hypothesising, researching and investigating'—the children will be provided with an opportunity to manipulate objects (their paper aeroplanes) and experiment with cause and effect, trial and error, and motion as they try to get their aeroplanes to fly and land in the number target. They will also be encouraged to use reflective thinking to consider why things happened and what can be learnt from this experience.
- 'Children resource their own learning through connecting with people, place, technologies and natural and processed materials'—the children will manipulate resources to assemble, invent, and construct their paper aeroplanes, which will allow them to explore ideas and theories.

In 'Learning Outcome 5: Children are effective communicators', the children will be working towards the following areas:

- 'Children interact verbally and non-verbally with others for a range of purposes', through answering open-ended questions and taking part in group discussions with their peers and educator. The children will also be given the opportunity to demonstrate an increasing understanding of measurement and number using vocabulary to describe size, length, and names of numbers.

Requirements

This experience will be set up during outside play on the soft fall, where the children will be able to fly their aeroplanes freely as well as have the opportunity to see real aeroplanes in motion, flying through the sky.

Materials required for this experience will be:

- laptop to show a short video to the children about how to make a paper aeroplane
- coloured paper for the children to make their aeroplane
- coloured pencils/markers for the children to decorate their aeroplane
- tape measure for the children to measure the distance their aeroplane flew
- masking tape to make the number target
- sheet of cardboard to keep record of the children's scores from the number target.

The experience set up is quite simple. All the materials that the children will need to create their paper aeroplanes, such as the coloured paper, pencils, and markers, will be located on the floor outside the area where the experience will be undertaken. The number target will be allocated next to the creating area and will need to be set up and created using the masking tape before the experience to ensure a smooth transition once the children have completed their aeroplanes. The laptop with the YouTube video will be preloaded and located on the floor ready for the children to watch and follow.

Procedure

This learning experience will be programmed to be flexible, which will allow for more child input and guidance. By setting out the learning experience in this manner, as their educator I am encouraging the children to put forth their own ideas and extend knowledge, and I am giving them the ability and opportunity to lead the experience in directions that spark their curiosity and interest.

To get the children's interest, I will approach them during free time play and initiate a conversation about transport such as fast cars and aeroplanes, which are subjects of interest to them—as I have learnt through my observations over the last two weeks. Once I have their attention I will then begin to use open-ended questions to help instigate an interest in discovery, which will be the catalyst to introducing the experience to the children.

When it is time for the experience to begin, I will direct the children over to the pre-set-up area and once again instigate a discussion about how they think we can make our aeroplanes. I will make use of open-ended questions that fall under the category of 'exemplify and specialise'—for example, 'What makes an aeroplane fly?' and 'What components will we need to make sure our aeroplane has so it will work?'

After the children have expressed their ideas and have an understanding of what they may need to do to make their aeroplane, I will then show them a quick video of how to make a paper aeroplane that will fly. Through the use of this technology, the children will be able to see the process and steps involved when making a paper aeroplane, which will help them acquire a holistic view. Once the children have watched the video once and have had the opportunity to ask any questions or expressed any ideas they may have, it will then be time for them to start to construct their own paper aeroplane. During this stage of the learning experience, the video will be playing in the background and, as the educator, I will be demonstrating and providing guidance and support where necessary. Through the use of demonstration, I will be able to support the children's learning by breaking the task into small sequential steps, which will allow time for them to observe intently and 'listen in' as the sequence of activities is presented. This stage of the experience is also a great time to engage the children in a discussion about the mathematical content area of geometry, such as the key concepts of shapes and spatial sense, part/whole, and patterns and relationships.

When the construction and decoration of their aeroplanes is finished, the children will then begin to test out their aeroplanes by taking turns at flying them and trying to get them to land on the different levels of the numerical target situated on the floor. At this time, the children will be encouraged to attempt to write down the target number on which their aeroplane landed—the children will write their numbers on a big piece of cardboard. After their second turn, with guidance and help where necessary, the children will be given the opportunity to interact with the concept of addition as they try to add their two scores together.

Because this experience is flexible in character and planned to be undertaken during outside play, there will be no need to end the learning experience as long as the children are still enjoying it. Teacher involvement can slowly decrease, which will allow for the children to take charge and continue the experience in a way that intrigues them. This will allow them to experiment and take on a more constructive approach to their learning. As this experience requires only minimal resources, the pack-away period will simply involve putting away the pencils and decorations used to decorate the aeroplanes (which could be done once the children had completed their decorations).

Plan for review

Throughout this experience it will be crucial to make note of what worked and what didn't work in order to ensure that the children are provided with the appropriate learning opportunities. At the end of the learning experience, I will write down a personal reflection about how I feel the experience went. In this personal reflection, I will make note of the effectiveness of the teaching strategies I used and come up with alternative methods to areas that didn't work. Another method I will use to review and evaluate this experience is to conduct a personal learning story on each child to see what they learnt from this experience. This will help me to see if the experience was appropriate and if I was able to facilitate the learning areas on which the experience was designed to focus.

(Source: Stephanie Pappas)

SYMMETRY IN 3D SPACE

Symmetry can also be applied to 3D shapes—though, conceptually, this is getting quite tricky for young children. Jorgensen and Dole (2011) provide the following useful example:

> Imagine cutting a shape in half. Do you have two identical shapes? If yes, then the cut that yielded the shape to mirror itself is called a plane of symmetry. Consider a solid figure such as a cylinder—how many planes of symmetry does it have? Since circles have an infinite number of lines of symmetry, cylinders have an infinite number of planes of symmetry (pp. 358–9).

FIGURE 13.8 **a** Reflection ('flip') (horizontal) and **b** Reflection ('flip') (vertical)

Although symmetry in 3D space is most likely to be met by children in the later primary school years, it is important to be mindful that the foundational knowledge that contributes to this concept is very much developed in the early childhood years—as such, do not be afraid to begin explorations of concepts such as symmetry with our young mathematicians!

Transformations

Transformation and symmetry are closely related concepts. **Transformation** refers to the alteration of a shape in some way. Alterations include changing the shape's position in space (congruent transformation), its size (projective transformation), or its features (topological transformation) (Jorgensen & Dole, 2011).

Transformation refers to the alteration of a shape.

Congruent transformations are those that alter the position of a shape but leave the shape itself unchanged. There are three types of congruent transformation: reflection, translation, and rotation—or, as they are more commonly known, flip, slide, and turn. Examples of each of these transformations are shown in Figures 13.8a–b, 13.9, and 13.10.

Congruent transformations are those that alter the position of a shape.

CURRICULUM CONNECTIONS

Australian Curriculum: Mathematics

Year two:

Investigate the effect of one-step slides and flips with and without digital technologies (ACMMG045)

- Understanding that objects can be moved but changing position does not alter an object's size or features.

  Identify and describe half and quarter turns (ACMMG046)
- Predicting and reproducing a pattern based around half and quarter turns of a shape and sketching the next element in the pattern.

(ACARA, 2022)

FIGURE 13.9 Translation ('slide')

FIGURE 13.10 Rotation ('turn')

Projective transformations are enlargements or reductions.

Topological transformations involve stretching or bending.

Projective transformations are enlargements or reductions of the original object (Jorgensen & Dole, 2011)—or, put simply, they are transformations that make the object bigger or smaller, while maintaining the original shape. Examples are shown in Figures 13.11 and 13.12.

Finally, **topological transformations** involve stretching or bending an object. Properties such as the lengths of lines, sizes of angles, and straightness are ignored (Jorgensen & Dole, 2011). Objects that have been topologically transformed are often described as 'skewed'. An example can be seen in Figure 13.13.

FIGURE 13.11 Enlargement

FIGURE 13.12 Reduction

FIGURE 13.13 Stretching an image

Can you think of some everyday activities that allow children to explore transformations? What sorts of resource might be useful for exploring projective and topological transformations?

Tessellation

Tessellation means fitting together shapes without gaps or overlaps.

Tessellation is a spatial concept that is easy to appreciate in everyday life—it is evident in architectural features such as tiling, paving, and brickwork, in geometric designs on fabric and clothing, and it also occurs in the natural world, to give just a few examples. **Tessellation** basically means fitting together shapes without gaps or overlaps, as seen in Figure 13.14.

FIGURE 13.14 Tessellated designs

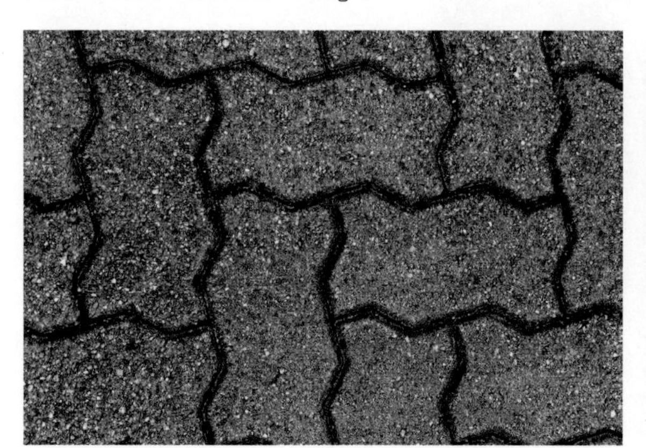

An individual shape is said to tessellate if it can be used repeatedly to cover an area with no gaps or overlaps. For example, a hexagon will tessellate, while a circle will not (Figure 13.15).

FIGURE 13.15 Examples of tessellating and non-tessellating shapes

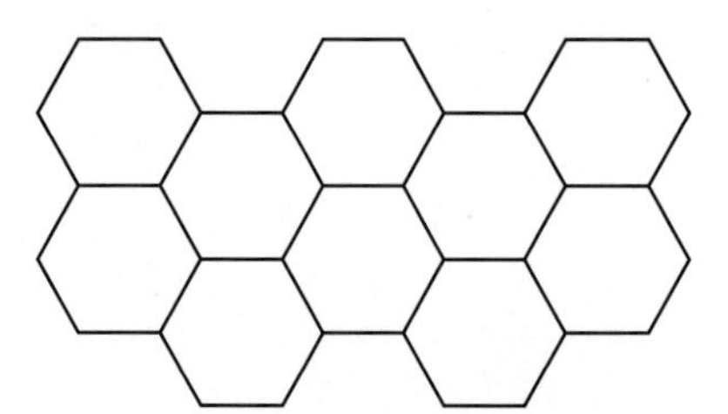

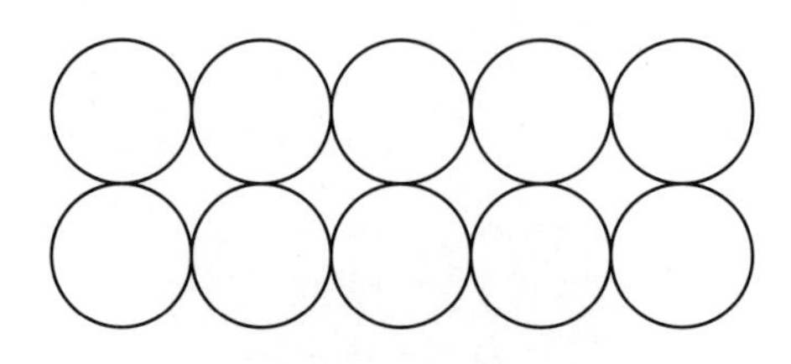

One of the simplest ways of introducing children to the concept of tessellation is to ask them to complete puzzles with interlocking pieces. This will help to reinforce the ideas of 'fitting together' and 'no gaps'. Creating mosaics and collages will also help children to explore these concepts.

Position and direction

The understanding of positional or locational language is a pivotal element in the development of children's spatial awareness. From a very young age, children are introduced to the language of position in meaningful ways—your bag goes *above* the shelf, the blocks are *below* the books, put your fork *beside* your plate. Understanding location of self and other objects is key to developing spatial perception (Siemon et al., 2011).

Positional language and concepts also relate specifically to direction—that is, developing understanding of north, south, east, west, and so forth. These terms are given meaning through everyday conversations, such as 'I live in East Albury', or 'We had to drive into the western sun on the way home'. Although children may not develop a good grasp of these specific directions until their primary school years, it is important to use these terms in everyday conversation so that when children encounter them in a mathematical context such as mapping, the terms already have some meaning for them.

CURRICULUM CONNECTIONS

Early Years Learning Framework

Outcome 4: Children are confident and involved learners

Children resource their own learning through connecting with people, place, technologies and natural and processed materials.

This is evident when children:

- Use their senses to explore natural and built environments.

Educators promote this learning when they:

- Provide sensory and exploratory experiences with natural and processed materials
- Provide experiences that involve children in the broader community and environment beyond the early childhood setting.

(DET, 2019, p. 40)

CURRICULUM CONNECTIONS

Australian Curriculum: Mathematics

Foundation year:

Describe position and movement (ACMMG010)

- Interpreting the everyday language of location and direction, such as 'between', 'near', 'next to', 'forward', 'toward'
- Following and giving simple directions to guide a friend around an obstacle path and vice versa.

Year one:

Give and follow directions to familiar locations (ACMMG023)

- Understanding that people need to give and follow directions to and from a place, and that this involves turns, direction and distance
- Understanding the meaning and importance of words such as 'clockwise', 'anticlockwise', 'forward' and 'under' when giving and following directions
- Interpreting and following directions around familiar locations.

(ACARA, 2022)

Mapping

Mapping is a mathematical activity that draws together many of the big ideas that have been presented in this chapter. In essence, a map is a representation of a space and its characteristics—and the creation of such a representation involves drawing upon knowledge of shape, position, direction, and relationships between spatial objects. For young children, mapping provides a meaningful context in which to apply these sorts of understandings. Research has shown that even three-year-old children have the capability to build simple maps (Clements, 1999a). Children should be encouraged to create maps of meaningful spaces—these may be real places such as the classroom or their bedroom, or they may be based on places with which they are familiar from storybooks or movies. Children should begin by constructing their own maps, making decisions about features they wish to include. Young children's maps typically do not correspond with 'reality' but rather express how they use their understanding of relative position to depict spaces and routes (Geist, 2009). Then, working from this basis, children can gradually be introduced to mapping conventions and standard features, such as a key and a scale. The main thing for children to recognise is the importance of representing the spatial relationships between the items within the map.

Learning Experience Plan 13.2

'We're going on an animal hunt'

Experience information

Children involved: Sarah, Toby, and Linkin.
Ages: All three children are five years of age.

Learning focus

The mathematical processes I am hoping to foster in this learning experience are designing, counting, and measuring. I will encourage the children to design a map that includes different locations within a space (outdoor environment) and then to *count* and *measure* the space within the designated area to recognise where each location is and then record this on their map. In this learning experience, the technological concepts and processes that will be evident will be those of drawing and map creating. Each map will be designed with an individual purpose to help each child achieve and carry out this learning experience. Through the experience, I am hoping to facilitate the children's awareness and understanding of space. In particular, the children's knowledge of spatial sense and spatial visualisation will be applied as the children engage in drawing and creating a map and visualising aspects within the outdoor environment to incorporate in their map.

It is clear that Sarah, Toby, and Linkin have all taken an interest in the plastic animals we introduced to the construction area inside and that they enjoy playing together as a group. The children have enjoyed characterising the animals, building homes and environments for them, and caring for them. Sarah, Toby, and Linkin often need extra encouragement to try new activities or to play in other environments away from the block area. My aim is to introduce the children to a new activity while fostering their interest in animals, and to facilitate new understandings of mathematical concepts and processes.

From my observations, Sarah, Toby, and Linkin have clearly identified their ability to work together collaboratively on group projects via different forms of interaction. Each child's social and language development is apparent in these collaborations. The children have also demonstrated their cognitive development and their ability to construct buildings and objects (using blocks) to suit space and play needs. Sarah and Toby have recently engaged in dramatic play as they characterised each animal they played with. Linkin has also characterised the tree he added to his play and recalled recent events at which he was present by saying that it was 'very windy'.

EYLF outcomes

The children within this group learning experience will be working towards EYLF Outcomes 4.2 and 4.4. The children will work towards these outcomes together and individually as they connect with the *technology* (map) in this experience to resource their own learning. The children will use important dispositions such as *cooperation*, *creativity*, and *imagination* as they go on an animal hunt together.

Requirements

This learning experience will be implemented in the outdoor environment of preschool, as this environment has many different components to foster the experience.

Materials required for the experience include:

- a sample map that I will create
- three clipboards and paper
- pencils or crayons
- three baskets to collect the animals once they are found.

We will need to be situated at a table or utilise a space on the veranda mat, providing the children with appropriate settings to draw their maps.

Procedure

First, I will need to ask Sarah, Toby, and Linkin if they would like to come and join in on this experience with me. As Sarah, Toby, and Linkin often play together, I will aim to approach the children as a group. Alternatively, I will group the children together in a quiet space within preschool to ask them if they would like to be involved in this learning experience and to inform them about what we will be doing if they decide to take part. The following sample sentences are examples of what I might say to capture the children's attention:

- 'Where are my friends Sarah, Toby, and Linkin? Could you please come and join me in book corner—I have something exciting that I want to do with you three, so come and listen to what it is!'
- 'Excuse me Sarah, Toby, and Linkin, when you are finished playing in this space, could you please come and find me. I need to go on an animal hunt and I would like you to help me find some animals that are outside hiding.'

As an educator, I will be involved within this experience from start to finish by being responsive to the children and also by supporting and building on their learning through social constructivism. My initial role is to gain the children's attention and explain what we will be doing. I need also to respect the children's ideas and thoughts about this learning experience to ensure that the experience fosters the children's curiosity and commitment. This will ensure that the children are motivated to participate and that the experience will have a positive impact on their learning. My role as the educator in this situation is also to allow time for the children to invest in the experience by letting them investigate, experiment, and explore.

If this experience needs to be ended or redirected to another area of the centre, I will need to gain the children's attention quickly and effectively. I will sing the centre's grouping song to gain the children's attention and redirect the play to another area, as modelled in the following examples:

- Example 1: 'Now that we have collected all the animals, let's go to the block area and build them all a house. How will we fit all of these animals inside a house?'
- Example 2: 'Wow, look at all those animals we have found! Sarah, Toby, and Linkin, what animal is your favourite? Let's go to the drawing table and draw a picture of our favourite animals. I am going to draw two big dogs and a small puppy.'

The children's interests and needs on the day will determine when we will pack away this learning experience. If more children wish to join in on the experience, we will leave all the resources and materials out so that children can access them easily and engage in the experience. However, if it is pack-up time at the centre, I would encourage Sarah, Toby, and Linkin to collect all of the animals and materials we have used within the experience and pack them away inside. I would also encourage the children to take responsibility for their maps and put them in a safe space inside the centre to take home if they wish.

Plan for review

- Throughout this learning experience, I will need to be aware of the children's reactions and statements to gauge if the children are enjoying the experience.
- By continually asking the children questions throughout this experience, I will be able to identify which elements of the experience worked well, and which didn't.
- Taking photos throughout this experience will help me reflect on its design and outcomes.
- I need to watch how the children engage in this experience so that I will be able to understand what they found hard and what they found easy—I will then be in a position to determine whether the children were challenged and if this experience was appropriate.
- I need to ask the children if they enjoyed this experience, what they didn't like, and what they would like to do next time.
- At the end of this experience, I will need to reflect and document my feelings and thoughts as an educator to determine whether my role was appropriate.

(Source: Elizabeth Bowden)

CURRICULUM CONNECTIONS

Early Years Learning Framework

Outcome 4: children are confident and involved learners

Children develop a range of skills and processes such as problem solving, inquiry, experimentation, hypothesising, researching and investigating.

This is evident when children:

- Create and use representation to organise, record and communicate mathematical ideas and concepts
- Explore their environment.

Educators promote this learning when they:

- Recognise mathematical understandings that children bring to learning and build on these in ways that are relevant to each child
- Encourage children to make their ideas and theories visible to others.

Children resource their own learning through connecting with people, place, technologies and natural and processed materials.

This is evident when children:

- Use their senses to explore natural and built environments.

Educators promote this learning when they:

- Provide experiences that involve children in the broader community and environment beyond the early childhood setting
- Provide resources that encourage children to represent their thinking.

(DET, 2019, pp. 38, 40)

CURRICULUM CONNECTIONS

Australian Curriculum: Mathematics

Year two:

Interpret simple maps of familiar locations and identify the relative positions of key features (ACMMG044)

- Understanding that we use representations of objects and their positions, such as on maps, to allow us to receive and give directions and to describe place
- Constructing arrangements of objects from a set of directions.

(ACARA, 2022)

Chapter summary

In this chapter I have presented the space and geometry concepts that can be explored in early childhood mathematics education. Collectively, facility with space and geometry concepts and skills forms the basis for spatial sense, grounding an ability to engage in geometric reasoning. Both spatial sense and geometric reasoning are critical for other domains of mathematical learning, such as patterning and measuring. It is essential that children are given a variety of opportunities to develop spatial and geometric knowledge through exploration of everyday objects and environments.

The specific topics we covered in this chapter were:

- spatial reasoning
- spatial orientation
- spatial visualisation

FIGURE 13.16 Exploring space

(Source: Amy MacDonald)

- van Hiele levels of geometric thinking
- shapes
- 2D and 3D shapes
- attributes of shapes
- lines
- symmetry
- transformations
- tessellation
- position and direction
- mapping.

FOR FURTHER REFLECTION

Consider Figure 13.16 of a photo taken in an early childhood classroom.

1 What everyday opportunities for exploring space and geometry exist in this type of environment?

2 How do everyday environments help to foster young children's spatial sense and geometric reasoning?

FURTHER READING

Jorgensen, R., & Dole, S. (2011). Patterns and algebra. In *Teaching mathematics in primary schools* (2nd ed., pp. 258–74). Crows Nest, NSW: Allen & Unwin.

Knaus, M. (2013). Geometry—spatial awareness and shape. In *Maths is all around you: Developing mathematical concepts in the early years* (pp. 61–76). Albert Park, VIC: Teaching Solutions.

Knaus, M. (2013). Pattern. In *Maths is all around you: Developing mathematical concepts in the early years* (pp. 22–32). Albert Park, VIC: Teaching Solutions.

Montague-Smith, A., & Price, A.J. (2012). Pattern. In *Mathematics in early years education* (3rd ed., pp. 83–114). New York: Routledge.

Montague-Smith, A., & Price, A.J. (2012). Shape and space. In *Mathematics in early years education* (3rd ed., pp. 115–44). New York: Routledge.

14 Measurement

Learn more
Go online to see a video from Paige Lee explaining the key messages of this chapter.

CHAPTER OVERVIEW

This chapter will explore the foundational measurement concepts that are developed in the early childhood years, present examples of these concepts, and provide example Learning Experience Plans that show how measurement can be explored with young children.

In this chapter, you will learn about the following topics:

- emergent and proficient measurement
- measurable attributes
- length
- area
- volume and capacity
- mass and weight
- time
- temperature
- value (money)
- comparison
- units.

KEY TERMS

- Accumulation of distance
- Area
- Area array
- Capacity
- Comparison
- Conservation of length
- Decomposing
- Depth
- Duration
- Emergent measurement
- Formal units
- Height
- Informal units
- Length
- Mass
- Measurement
- Partitioning
- Proficient measurement
- Relation to number
- Sequence
- Superimposing
- Temperature
- Time
- Transitivity
- Unit
- Unit iteration
- Value
- Volume
- Weight
- Width

AMY MACDONALD

Introduction

Measurement is the assignment of a numerical value to an attribute of an object or event.

This chapter describes the interconnected elements that contribute to a child's understanding of measurement. Jorgensen and Dole (2011, p. 275) define **measurement** as 'the assignment of a numerical value to an attribute of an object or event'. Measurement also involves dimensions of continuity, comparison, and order, which allow objects and events that may not be separately countable to be arranged or categorised and compared.

'Measurement' is a mathematical knowledge domain encompassing a range of concepts; and, importantly, it is also a *process* (i.e. 'measuring') whereby the concepts involved interact in such a way that a measurement of—an assignment of value to—an object or event can be made. Relationships between measurement processes and measurement concepts contribute to a child's emergent understanding of measurement; and once this understanding is consolidated, they contribute to the development of more proficient applications of measurement. Considering prior research on children's measurement knowledge, and the shift in curriculum materials towards combining different elements of mathematical understanding, this chapter takes an integrative approach to exploring children's emerging measurement understandings.

Research has shown that understandings of measurement begin to develop in the early childhood years. Young children know that attributes such as mass and length exist, although they may not be able to quantify or measure them accurately (Clements & Stephan, 2004). As Clements and Stephan explain, 'even 3-year-olds know that if they have some clay and then are given more clay, they have more than they did before' (p. 300).

By about four to five years of age, most children begin to make progress in reasoning about and measuring quantities, and they learn to use words that represent quantity of a certain attribute. Children then learn to compare two objects directly and recognise equality or inequality. Children also learn to measure by connecting number to quantity and identifying a unit of measure. In short, learning about measurement and the measuring process involves the development of three key concepts:

1. Objects and events have attributes that can be measured.
2. Measurement can be used to compare objects and events.
3. Formal and informal units can be used to measure objects and events.

Collectively, these three key concepts represent components of the measuring process—namely, identifying measurable attributes of objects and events, comparing objects and events on the basis of these attributes, and using units (formal or informal) to measure objects and events.

These three concepts, and the areas of knowledge associated with them, will be explored in this chapter.

CURRICULUM CONNECTIONS

Early Years Learning Framework

Outcome 5: Children are effective communicators

Children interact verbally and non-verbally with others for a range of purposes.

This is evident when children:

- Demonstrate an increasing understanding of measurement and number using vocabulary to describe size, length, volume, capacity and names of numbers
- Use language to communicate thinking about quantities to describe attributes of objects and collections, and to explain mathematical ideas.

Educators promote this learning when they:

- Listen to and respond to children's approximations of words
- Model language and encourage children to express themselves through language in a range of contexts and for a range of purposes
- Include real-life resources to promote children's use of mathematical language.

Children begin to understand how symbols and pattern systems work.

This is evident when children:

- Begin to be aware of the relationships between oral, written and visual representations.

Educators promote this learning when they:

- Engage children in discussions about symbol systems, for example, letters, numbers, time, money and musical notation.
- Encourage children to develop their own symbol systems and provide them with opportunities to explore culturally constructed symbol systems.

(DET, 2019, pp. 43, 46)

Emergent and proficient measurement

It could be said that the development of children's measurement understandings are divisible into two levels, these being *emergent measurement* and *proficient measurement*. **Emergent measurement** encourages children to develop an understanding of measurement by using it for their own purposes, talking about their measurement ideas, representing measurement processes in ways that make sense to them, and becoming more aware of their own measurement thinking (Whitebread, 2005). On the other hand, **proficient measurement** requires comprehension of measurement concepts, operations, and relations; skills in carrying out procedures flexibly, accurately, efficiently, and appropriately; ability to formulate, represent, and solve problems; and capacity for logical thought, reflection, explanation, and justification (Kilpatrick, Swafford & Findell, 2001).

Emergent measurement involves children using measurement for their own purposes in meaningful ways.

Proficient measurement involves children comprehending measurement concepts, operations, and relations.

What examples of *emergent* and *proficient* measurement have you observed in your interactions with young children? How can you determine the difference between the two?

While it's important to capitalise on children's natural interactions with measurement, as educators we have to be very mindful of our role in advancing conceptual understandings from emergent to proficient. We as educators must be proficient *ourselves* in understanding and using measurement in flexible and appropriate ways. This means we can't just provide children with lots of interesting and stimulating experiences and *hope* they'll learn; we have to model, question, and use appropriate mathematical language *with intention* as a means of facilitating understandings of key measurement concepts. These key concepts will now be explored in greater depth.

Measurable attributes

The first step in understanding measurement and using measuring processes is recognising that all objects and events have measurable attributes. Measurable attributes of objects include:

- length
- area
- volume and capacity
- weight and mass
- time
- temperature
- cost (associated with money).

Each of these measurable attributes—and the concepts associated with these attributes—will now be explored in greater depth.

LENGTH

Length
refers to how long something is, along a single plane or through two dimensions.

Height
refers to how high something is, along a single plane or through two dimensions.

Width
refers to how wide something is, along a single plane or through two dimensions.

Depth
refers to how deep something is, along a single plane or through two dimensions.

Length is thought to be the most easily understood of the measurement concepts and refers to how long something is, whether along a single plane or through two dimensions (Reys, Lindquist, Lambdin, & Smith, 2007; Jorgensen & Dole, 2011). It is usually one of the first measurement concepts that children encounter, and most children have little difficulty developing the concept of length and the language associated with it (Jorgensen & Dole, 2011). Length is a more generic term associated with all linear measurement, but the terms **height**, **width** and **depth** are also used to refer to specific linear attributes along a single plane or through two dimensions, as illustrated in Figure 14.1.

According to Stephan and Clements (2003), linear measurement involves six important concepts:

1. partitioning
2. unit iteration
3. transitivity
4. conservation
5. accumulation of distance
6. relation to number.

FIGURE 14.1 Height, width, depth

Partitioning
is dividing an object into equal-sized units.

What measurement language might children use when playing on play equipment or at a playground?

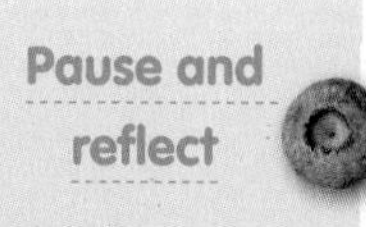

Partitioning is the mental activity of dividing an object into equal-sized units. This idea is not immediately obvious to children, as it involves conceptualising the object as something that can be divided (partitioned) *before* it is physically measured (Clements & Stephan, 2004).

Unit iteration is the ability to think of the length of a unit as part of the length of the object being measured and to place the unit end to end repeatedly along the length of the object (Kamii & Clark, 1997).

Transitivity is the understanding that if the length of object *A* is equal to (or greater/less than) the length of object *B* and object *B* is the same length as (or greater/less than) object *C*, then object *A* is the same length as (or greater/less than) object *C* (Clements & Stephan, 2004). A child who can reason transitively can take a third or middle item as a referent by which to compare the heights or lengths of other objects.

Conservation of length is the understanding that as an object is moved, its length does not change (Piaget, 1969a). Although researchers agree that conservation is essential for a complete understanding of measurement, it has been cautioned that children do not necessarily need to develop transitivity and conservation before they can learn some measurement ideas (Boulton-Lewis, 1987; Clements, 1999b; Hiebert, 1981).

Accumulation of distance is the understanding that as you iterate a unit along the length of an object and count the iteration, the number words signify the space covered by all units counted up to that point (Clements & Stephan, 2004).

Finally, **relation to number** requires children to reorganise their understanding from the counting of discrete units to the measure of continuous units. This involves knowing that when counting discrete units, the last number in the counting sequence represents the total number of units—the measurement of the object.

Unit iteration
is thinking of the length of a unit as part of the length of an object, and placing that unit end to end along the object.

Transitivity
is knowing that if *A* is as long as *B*, and *B* is as long as *C*, then *A* is also the same length as *C*.

Conservation of length
is knowing that if an object is moved, its length does not change.

Accumulation of distance
is knowing that as you iterate the unit and count the units, the numbers represent the space covered by the units.

Relation to number
is knowing that the last number when counting units represents the measurement of the object.

Learning Experience Plan 14.1

'Pit stop'

Experience information

Four focus children:

J, 5 years 8 months
C, 4 years 9 months
M, 4 years 11 months
J, 4 years 10 months.

Learning focus

The learning objective for this experience will be to extend children's physical scientific and mathematical understanding. The experience will allow the children to explore machines and measurement through creating vehicles that can roll, measuring how far each vehicle can roll, and thinking about why it did or didn't roll far. The key to this learning experience lies in the children's interest in building and constructing, and comparing measurements. The children have shown a sound understanding of measurement and machines, and with the support of technology the main outcomes will be to broaden children's learning beyond basic construction, and to deepen their understanding of the different attributes of machines and how they work.

The technology used in the experience will support the children's ability to make reasoned predictions by enabling each child to design and construct their own vehicle and then to observe the performance of their design and make changes accordingly. Technology can be used to support children's learning in achieving other developmental milestones such as cognitive, physical, and social and emotional skills, and it can contribute to sustained and playful engagement in learning.

Another outcome of this learning experience is to create an environment in which the children, as naturally 'social beings', enjoy engaging in group work and are prepared to recognise and celebrate each other's achievements through theory building. The role of the educator will include social help, supporting the children as they develop problem-solving and collaborative skills, and encouraging them to work together to share and discuss ideas and listen to new perspectives. This learning experience is age appropriate for this group of children in the way that it inspires both individuality and collaboration, is thought provoking, and is based solely on child-instigated investigations and interests.

EYLF outcomes

EYLF Outcome 4: Children are confident and involved learners

Children develop a range of skills and processes such as problem solving, inquiry, experimentation, hypothesising, researching and investigating.

- The children will practise a number of thinking strategies while participating in the experience—they will solve problems that are at appropriate levels of challenge and they will be encouraged to explore, experiment, and take appropriate risks in their learning.
- The children will manipulate objects and experiment with cause and effect, trial and error, and motion during the experience—activities that will encourage investigation and problem solving.
- The educator will model mathematical and scientific language, reasoning, predicting, and reflecting processes, thereby prompting the children to reflect on and consider why things

happen and what can be learnt from the activities undertaken in the learning experience. This should enhance the children's ability to contribute constructively to mathematical discussions and arguments.

Children develop dispositions for learning such as curiosity, cooperation, confidence, creativity, commitment, enthusiasm, persistence, imagination, and reflexivity.

- Since the learning experience will be open-ended and flexible, it will allow the children to be creative, and to take control, investigate, imagine, and explore.
- Children come to learning experiences with enthusiasm, energy, and concentration when their interests and ideas are listened to and their desire to extend on them has been recognised.

Requirements

The setting for this learning experience will be within the indoor environment on a large mat, where there is sufficient room for the children to sit and create their own vehicles with the Dacta Duplo, and to set up a racing track beside them.

The materials that will be included are:

- Dacta Duplo
- basket of blocks
- colourful markers
- masking tape
- large floor area
- 'Pit Stop' sign
- 'Pit Garages' sign
- 'Race Track' sign
- large cardboard chart.

The construction area will have a sign on the wall marked 'Pit Stop'. I will place masking tape on the mat, which will be designed to look like a race track with no finish line (the track will also be labelled with a 'Race Track' sign). Alongside the race track there will be a basket of wooden blocks for the children to measure the distance each vehicle travels.

Procedure

This week in town is the Bathurst 1000 race week. As a preschool we walk into town and watch the large trucks in the truck parade as they drive through town to 'the Mountain'.

The learning experience will channel the excitement that the children experience during the lead-up to the Bathurst 1000. The children love to build together, and their sense of belonging will be strengthened, and their mutual interests sparked, by the opportunity to create their vehicles, to make adjustments to them where they feel necessary, and to measure and track each vehicle's performance. The children will be able to create their own motor race, take ownership of the creative process involved, and build confidence from the tasks that they complete.

I will introduce to the group the pit-stop and race-track areas, explaining that these areas together form the preschool's very own Mount Panorama. When the children have created a vehicle in the pit stop, I will encourage them to see how far their vehicle can roll with one push on the preschool race track. When the vehicle comes to a stop, the children can line the blocks up from the starting line to the point at which their vehicle lies. The children can measure the distance that each vehicle has travelled by rationally counting the number of blocks needed to cover the passage of the vehicle's journey. As a group, we will write on a chart the number of blocks that each vehicle has travelled. The aim of the chart is to link the number of blocks counted to the written numeral, to compare different lengths, to use different terms—'more, longer, less', etc.—and to ask open-ended questions to provoke the children's learning. These open-ended questions encourage children to organise and express comparisons and measurements: 'What did you see?' 'What happened first?' 'Tell us what you think about ...' 'What do you think will happen next?'

The pit stop will be open for each child to come and go as they please to make adjustments to their vehicle, explore the balance and design of their vehicle, to change the design completely or to make something else altogether. There will be a shelf available, labelled 'Pit Garages', for any child wanting to undertake another activity but also wanting to keep their creation intact. The garages will also keep the children's creations safe when it is time for pack away.

Plan for review

I will determine if the learning experience is appropriate by being aware of, and documenting, children's engagement in the activity and their responses to the open-ended questions. I will ensure that the strategies used in aiding intentional teaching practices encourage the children in their interests, and engage them as confident and capable scientists and mathematicians; I will maintain a balance between intentional teaching practices and child-initiated learning. I will create a plan for reviewing the experience that will assist me to extend and challenge the children's learning as appropriate. I will be watching for any unstructured learning in which the children engage and will support them by following up on their discoveries and ideas.

(Source: Julia Alexander)

AREA

Area
refers to the amount of space contained within a 2D shape.

Area refers to the amount of space contained within a two-dimensional shape. Research has suggested that a good understanding of linear measurement is a prerequisite to a good understanding of area measurement, as both linear and area measurement rely on many of the same ideas related to units (Izsák, 2005). These ideas include the following:

- that a relation holds between the unit of measurement and the attribute being measured
- that a fixed unit can be iterated and that length or area can be partitioned into a number of equal-sized units
- that unlike units cannot be counted the same
- that measurement units should—without overlapping—cover or fill the attribute being measured
- that the size of the unit is inversely proportional to the measure of a quantity.

An **area array**
is formed when a unit is iterated in two directions.

Furthermore, the unit is iterated in two dimensions to create an **area array**, and this leads to multiplicative relationships involving the lengths of the sides (Curry & Outhred, 2005). An example of an array can be seen in Figure 14.2.

FIGURE 14.2 Making an array

(Source: Amy MacDonald)

Studies conducted by Lehrer (2003); Lehrer, Jenkins, and Osana (1998); and Stephan and Clements (2003) have found that children do not appear to develop these ideas in a predictable order; a fully developed understanding of area measurement requires the coordination of multiple ideas, and the process of coordination can vary from child to child (Izsák, 2005). Think about the non-metric forms of measurement that we regularly use when considering area measurement. Generally, children are introduced to the concept of area by **superimposing** areas and, later, by measuring areas with informal units. By using informal units that are familiar to children, an understanding of the structure of the units (an array) can be developed. Children must also develop an understanding that **decomposing** and rearranging shapes does not affect their area (Clements & Stephan, 2004).

Superimposing means laying one object on top of another object.

Decomposing means 'pulling apart' or 'deconstructing' an object.

VOLUME AND CAPACITY

The terms 'volume' and 'capacity' are often used interchangeably; however, these two terms have different (but related) meanings. **Volume** refers to the amount of space taken up by an object, while **capacity** refers to the amount an object can hold. The measurement of volume and capacity shares many features with linear and area measurement, such as the importance of unit iteration and the relation between measure and unit size (Curry & Outhred, 2005). Volume can be measured in two ways: the first method is to 'pack' the space with a three-dimensional array consisting of a two-dimensional array of units that is iterated in the third dimension; the second method is to 'fill' the space by iterating a fluid unit that takes the shape of the container (Curry & Outhred, 2005, p. 265).

Volume refers to the amount of space taken up by an object.

Capacity refers to the amount an object can hold.

The concepts and language of volume and capacity are extremely complex. Historically, researchers and theorists such as Piaget have argued that children do not understand volume, or how much space something takes up, before they are eleven or twelve years of age. This is because difficulties arise in making the transition from filling a space with concrete units to visualising and using the unit structure (Battista & Clements, 1996). However, it seems that some children do indeed develop an understanding of these concepts during the early childhood years, due to the many uses of volume and capacity in everyday environments—for example, water play, cooking, filling the car with petrol (Gifford, 2005). As such, relating learning experiences to children's informal knowledge and everyday contexts may help them to grasp volume and capacity concepts at a much younger age.

This is where water trays and sandpits—and time to play in them—become critically important. From quite simple observations about 'full', 'empty', 'overflowing' to creating and comparing referent objects and applying principles of transitivity, sandpits and water trays are very effective pedagogical tools. For example, by simply providing children with a variety of containers in the sandpit (Figure 14.3), we open up opportunities for conversations and explorations about the properties of containers, for instance:

- 'Which container is the biggest/smallest? How can you tell?'
- 'How many scoopfuls of sand will it take to fill the bucket?'
- 'Which container holds more sand? How can we find out?'

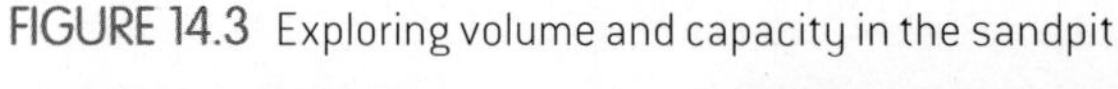

FIGURE 14.3 Exploring volume and capacity in the sandpit

(Source: Amy MacDonald)

'Exploring capacity using coloured water'

Learning Experience Plan 14.2

Experience information

Children: JB, SP, MH, KL

Ages: all aged 5 years

Learning focus

The learning focus of this experience is to extend the existing knowledge of four children (listed above) on measurement (in particular mass) by transferring this knowledge to capacity and volume. The focus of this experience is to expand the children's current understanding of more and less, and mass conversion, by allowing them to experiment, predict, and hypothesise about the capacity of different shaped/sized containers. The learning experience will introduce the children to non-standard and standard units of measurement. By using everyday language to facilitate children's understanding of a concept we can extend this learning opportunity by introducing more formal language associated with the concept. Similarly, by using simple non-standard measurements on the side of containers to extend the ideas of more and less, we are able to introduce a more formal method of measuring—i.e. measuring jugs—enabling children to extend their knowledge of measurement before we introduce standard units of volume.

I have chosen to extend the learning of this group of children, using capacity and volume to assess whether the children can transfer their existing knowledge of more and less to other contexts—i.e. from mass to capacity and capacity to volume. As the weather is warming up I decided on water play outside.

EYLF outcomes

- I will be asking the children to hypothesise which container holds more/less, and will be permitting them to express their own ideas and respond to ideas from others. By allowing them to feel supported in expressing their own ideas and considering the ideas of others, I will be supporting their sense of identity and giving them a setting in which to explore and learn, thereby stimulating their confidence—Outcome 1.
- When children have confidence to explore and learn, they become open to new challenges and become increasingly aware of the rights of others to have their own opinions and ideas. By involving all four children in discussions and predictions and challenging their existing knowledge, I will be developing their emerging autonomy and inter-dependence—Outcome 1.
- By encouraging the children to experiment with water and containers, and to work collaboratively with others, I will be allowing each to interact, constructively and with respect, as part of a group—Outcome 1.
- By planning an opportunity for these children to extend their knowledge, and by prompting them to investigate their ideas and concepts in a collaborative environment, I will be supporting their sense of belonging within a social group—Outcome 2.
- By permitting the children to express their own ideas and exposing them to different ideas (of the other children), I will be allowing them to respond to diverse perspectives with respect and to become socially responsible—Outcome 2.
- By building on and extending the children's existing knowledge of mass, and by acknowledging their efforts during the experience and celebrating their success, I will be supporting their social and emotional wellbeing—Outcome 3.

- By providing an active learning environment and giving the children opportunity to experiment and explore by transferring learning from one context to another, I will be developing their disposition to learning—Outcome 4.
- I will be questioning the children and creating an environment of wonder, and using play to foster the children's confidence to make predictions, hypothesise, investigate, and explore. I will be allowing children to be curious and enthusiastic learners—Outcome 4.
- By introducing mathematical language and symbols, and the tools of measurement, I will be developing additional skills and processes in mathematical concepts—Outcome 4.
- I will be sponsoring the transfer of existing knowledge from one setting to another: mass (play dough) to capacity and volume (water, containers, and measuring jugs)—Outcome 4.
- By encouraging the use of different shaped/sized containers and tools for measuring, I will be promoting the enhancement of the children's learning through technology—Outcome 4.
- I will be introducing mathematics and technology terminology, and will therefore be assisting the children to interact and use language for a range of purposes—Outcome 5.
- By using non-standard and standard measuring tools, I will be extending the children's use of symbols as communication aids—Outcome 5.

Requirements

Setting for the experience:

- a water trough and table set up outside, under the shade of the tree.

  Materials required:
- water trough
- hose to fill the water trough
- table
- various shaped, clear plastic bottles and jars (clear so that the boys can see the liquid inside). Fifteen vinegar, water, cordial, and soft-drink bottles in varying sizes will be used; and seven spread jars will be used
- green food colouring (water to be coloured so that it is visible through clear containers)
- marker to mark non-standard measuring marks on the side of some bottles and jars
- eight 1L measuring jugs with standard metric measures.

  Presentation of experience:
- The water trough and table will be set up at right angles to each other under the large tree.
- Prior to the children's arrival into care, the trough will be half filled with water by the hose and coloured with green food colouring.
- Six of the bottles and three of the jars will be marked with equal non-standard measures.
- Three of each bottles and jars will be filled with water from the trough and placed on the table.
- The remaining bottles and jars will also be placed on the table but will remain empty.

Procedure

- I will be at the water trough filling and emptying containers.
- If the children are not drawn to the water play, I will call them over.
- I will allow the children ample time to fill and empty containers on their own and take particular note of their vocabulary—introducing terminology more/less, full/empty, fill/empty.

- The term 'capacity' will be introduced—as the maximum amount of water with which a container can be filled.
- I will note whether the children are making the correlation between different shaped/sized containers holding the same/different quantities of water.
- I will ask the children what they think the numbers (or non-standard measures) on the sides of some bottles are—'What are these numbers on the side of the bottles?' 'How do you think they work?' 'What do they tell us?'
- Drawing their attention to the measuring jugs, I will ask—'What are these numbers on this jug? They look like the numbers on the bottle but seem to be different. Why do you think they are different? There is an "mL" next to these numbers and an "L" next to this number at the top: I wonder what they mean?'
- The term 'volume' will be introduced as the amount of space the water takes in a bottle or jar.
- To engage the children in a measuring experience, I will hold up two bottles and ask which one they think holds more water and note their replies. I will extend their thinking by asking why they think the designated bottle has the most water.
- My next question will be—'How do you think we can measure which bottle holds the most?'
- I will then ask the children to 'measure' the water in each of the bottles to test out their theories.
- The children will then be encouraged to experiment with the different sized containers and measuring tools independently.

  Concepts to be explored:
- more/less
- full/empty
- fill/empty
- capacity
- volume
- measuring
- standard/non-standard measures
- recognising numerals.

  Terminology to be introduced:
- capacity
- volume
- measure (ment)
- millilitres
- litres.

Plan for review

- I will assess the appropriateness of this learning experience by the engagement level of the children. This will be evident if the children experiment and explore the capacity of the various shaped/sized bottles and jars. Further evidence will be the engagement of the children in the experience when I move away and stop leading their play.
- My teaching strategies of posing questions to heighten their curiosity, experimentation, and exploration will be assessed by the children's achievement of (or failure to achieve) the learning outcomes of transferring existing knowledge of measurement (mass) to capacity and volume, and embracing the new terminology introduced during the experience.

- The learning and/or development facilitated, and the suitability and success of the learning plan, will be assessed by observing the boys children relationships between different shaped/sized bottles holding the same/different amount of water. This will be achieved by allowing the children to predict, hypothesise, and experiment—having a hands-on approach.
- My second learning focus is to introduce new terminology—this will be assessed for learning, suitability, and success by the children's capabilities to understand new terms and use them in appropriate contexts, especially when playing independently of my interactions.
- If the children do not engage in the experience, or do not engage with me and demonstrate curiosity about the relationship between different shaped/sized bottles holding the same/different amount of water, I will rethink my methods and interactions, re-evaluate my learning focus, and replan this experience.

(Source: Patsy Saul)

MASS AND WEIGHT

Mass
refers to the amount of matter in an object.

Weight
refers to the force that gravity exerts on an object.

Much like 'volume' and 'capacity', the terms 'mass' and 'weight' are often thought to have the same meaning; this, however, is not the case. **Mass** is defined as the amount of matter in an object, while **weight** refers to the force that gravity exerts on an object.

Mass and weight can be difficult concepts for young children to grasp because they cannot be seen. Young children's experience of mass will relate to specific experiences, including weighing people and moving heavy objects. These experiences need to be identified and built upon in order to develop children's understanding of mass. Because mass is invisible, children often relate mass to the visual attributes of the object being weighed. A common misconception is that larger things weigh more than smaller things; so to acquire an appropriate understanding of mass, children need to discuss large light things and small heavy things (Gifford, 2005). This is an aspect of mass that can easily be related to children's everyday experiences in a range of settings. For example, balloons, when filled with air, are really useful because they can take on a large shape and remain relatively light. They can also be used to demonstrate the different masses of different matter. For example, when filled to a specific size with water the balloon is heavier than when filled to the same size with air. Likewise, when filled with a particular gas, such as helium, the balloon is significantly lighter than when filled with air.

CURRICULUM CONNECTIONS

Australian Curriculum: Mathematics

Year two:

Compare masses of objects using balance scales (ACMMG038)

- Using balance scales to determine whether the mass of different objects is more, less or about the same, or to find out how many marbles are needed to balance a tub of margarine or a carton of milk.

(ACARA, 2022)

TIME

Time is a concept that is often quite difficult for children to learn due to its abstract nature. According to Piaget (1969b), in order to acquire the concept of time, children must grasp two important ideas: 1. there are a series of events that occur in a temporal order; and 2. between these events there are intervals whose duration must be appreciated.

Time
refers to the notion that events occur in a temporal order, and that events have duration.

In short, children must learn the attributes of **sequence** (the order of events) and **duration** (how long an event takes). In answer to the question 'What do we know about how young children begin to understand the concept of time?', Barnes (2006, p. 291) claims that 'we do know that mathematics learning builds on the curiosity and enthusiasm of children and grows naturally from their experiences (NCTM, 2000, p. 73). We also know that mathematics needs to be appropriately connected to the young child's world'. With this in mind, we must acknowledge that children develop an understanding of time by connecting it in ways that have meaning for them (Barnes, 2006). As Charlesworth (2005) explains, children relate time to three things:

Sequence
refers to the order in which events occur.

Duration
refers to how long an event takes.

1 personal experience
2 social activity
3 culture.

Thus, it is important for teachers to provide opportunities for children to develop time concepts in ways that are meaningful and personalised. Exploring children's developing understandings of time through representations may provide a crucial means by which to develop meaningful and personalised time concepts.

It makes sense, then, that we should let students investigate and express their daily activities through developing sequences—for example, time lines. Thus, a child might develop a time line for a whole day, a specific amount of time (i.e. lunch time), or something more immediate like dance steps or a Lego construction. It is also important to recognise that children are capable of representing time in different ways (for example, through pictorial images, or more traditional forms such as clock faces), and that these representations do not necessarily develop in a linear informal-to-formal fashion—that is, a child may know how to represent 3 o'clock on an analogue clock face before they are able to draw a sequence of events; or the inverse may also occur. Importantly, learning experiences about time need to be based on the child's own personalised experiences with, and understandings of, sequence and duration.

Educator reflection

My toddlers are very interested in their older family photos now. They are enjoying seeing just how much they have grown. I hear things like 'baby Jasper'. I wonder whether incorporating a measuring chart and more photos will assist with this interest. That way, children may develop a more visual appreciation of their growth over time.

(Source: Amy Urquhart)

Learn more
Go online to hear this Educator reflection being read.

CURRICULUM CONNECTIONS

Australian Curriculum: Mathematics

Foundation year:

Compare and order duration of events using everyday language of time (ACMMG007)

- Knowing and identifying the days of the week and linking specific days to familiar events
- Sequencing familiar events in time order.

  Connect days of the week to familiar events and actions (ACMMG008)
- Choosing events and actions that make connections with students' everyday family routines.

Year one:

Tell time to the half-hour (ACMMG020)

- Reading time on analogue and digital clocks and observing the characteristics of half-hour times.

  Describe duration using months, weeks, days and hours (ACMMG021)
- Describing the duration of familiar situations such as 'how long is it until we next come to school?'

Year two:

Tell time to the quarter-hour, using the language of 'past' and 'to' (ACMMG039)

- Describing the characteristics of quarter-past times on an analogue clock, and identifying that the small hand is pointing just past the number and the big hand is pointing to the three.

  Name and order months and seasons (ACMMG040)
- Investigating the seasons used by Aboriginal people, comparing them to those used in Western society and recognising the connection to weather patterns.

Use a calendar to identify the date and determine the number of days in each month (ACMMG041)

- Using calendars to locate specific information, such as finding a given date on a calendar and saying what day it is, and identifying personally or culturally specific days.

(ACARA, 2022)

TEMPERATURE

Temperature refers to the warmth or coldness of an object or substance.

Temperature can be defined as the measure of the warmth or coldness of an object or substance, with reference to a standard value. Despite the usefulness of this definition, it is very often the case that temperature is overlooked as a measurable attribute. However, it can be logically argued that temperature is in many ways no different from length, or mass, or any other commonly held measurable domain—temperature is an *attribute* that can be identified ('gosh it's hot today!'), that can be compared ('yeah, it's heaps hotter today than yesterday'), that can be measured both informally ('you could fry eggs on that concrete!') and formally ('the mercury soared above 40°'), and to which we assign formal measurement systems, units, and instruments.

It is important to recognise that children learn about, and experience, temperature as a concept in a range of ways in everyday life. The simple act of choosing clothing on the basis of the weather is an expression of temperature as a measurable concept. Children interact with temperature countless times throughout the day—for example, 'Put your jacket on, it's cold outside'; 'Would you like a cold drink?'; 'How long do we need to put the baked beans in the microwave?'

Temperature is an example of how mathematical learning can be highly contextualised and meaningful in everyday life. It also opens the door to a range of opportunities for connecting with other areas of mathematics in meaningful ways. For example, investigating weather patterns and charting temperatures are great ways of developing concepts of data collection and representation.

VALUE (MONEY)

Typically, textbooks and curriculum documents reserve money for inclusion in the 'number' domain, exploring money as an example of learning about number concepts such as decimals, place value and base 10 systems. However, money can also be considered a measurement system, applying the logic that things have a **value** that can be measured, and that dollars and cents (for example) are measurement units. Of course, money requires understanding of the aforementioned concepts—decimals, place value, base 10—but by viewing the application of money as the *measurement of value*, we are immediately ascribing a meaningful, real-world application to this concept. Indeed, the concept of money is one of the most visible mathematical ideas in children's everyday lives, with many opportunities to explore this measurement system in routine activities such as grocery shopping (Figure 14.4).

Value
can be taken to mean the measurement of an object's importance, worth, or usefulness.

FIGURE 14.4 Money in everyday life

(Source: Amy MacDonald)

What are some other everyday contexts in which children might learn about the measurement of value, i.e. money?

CURRICULUM CONNECTIONS

Australian Curriculum: Mathematics

Year one:

Recognise, describe and order Australian coins according to their value (ACMNA017)

- Showing that coins are different in other countries by comparing Asian coins to Australian coins
- Understanding that the value of Australian coins is not related to size
- Describing the features of coins that make it possible to identify them.

Year two:

Count and order small collections of Australian coins and notes according to their value (ACMNA034)

- Identifying equivalent values in collections of coins or notes, such as two five-cent coins having the same value as one 10-cent coin
- Counting collections of coins or notes to make up a particular value, such as that shown on a price tag.

(ACARA, 2022)

Comparison

Comparison involves making a judgment about the similarities or differences between two or more objects or events.

A step in the progression of young children's knowledge about measurement is developing an understanding that measurement can be used to compare objects. **Comparison** involves making a judgment about the similarities or differences between two or more objects or events. Understandings of comparison typically progress in the following manner:

1 understanding that the key idea is to compare like attributes
2 comparing objects directly
3 making multiple comparisons of objects.

Comparing like attributes requires knowledge of the measurable attributes of objects (i.e. length, area, mass), and an ability to identify the same attribute in different objects.

Comparing objects directly involves making judgments about two objects on the basis of a common attribute. Often this will involve placing two objects against one another to make a visual comparison, usually along a common baseline, such as in Figure 14.5. This process will enable a child to make reliable judgments about, for example, which of two objects is the taller or the shorter.

FIGURE 14.5 Direct comparison using a common baseline

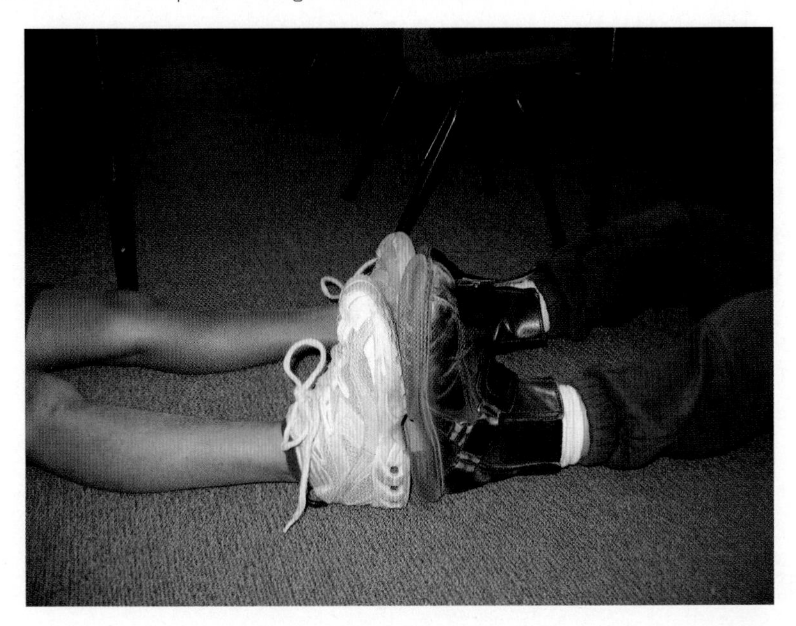

(Source: Amy MacDonald)

The making of multiple comparisons of objects is the most complex comparative process, and it requires children to compare more than two objects that are linked by some sort of common attribute or feature. This process may or may not involve a series of direct comparisons. The use of a common baseline may assist children in making determinations of this kind—such as which is the tallest or shortest member of the set, as shown in Figure 14.6.

FIGURE 14.6 Multiple comparisons along a common baseline

(Source: Amy MacDonald)

CURRICULUM CONNECTIONS

Australian Curriculum: Mathematics

Foundation year:

Use direct and indirect comparisons to decide which is longer, heavier or holds more, and explain reasoning in everyday language (ACMMG006)

- Comparing objects directly, by placing one object against another to determine which is longer or by pouring from one container into the other to see which one holds more
- Using suitable language associated with measurement attributes, such as 'tall' and 'taller', 'heavy' and 'heavier', 'holds more' and 'holds less'.

(ACARA, 2022)

Units

A unit is a quantity used as a standard of measurement.

Units are important for communicating measurements, as they enable a standardised and shared interpretation of the measure. A **unit** can be defined as a quantity used as a standard of measurement. Measuring as a process consists of 'identifying a unit of measure and *subdividing* (mentally and physically) the object by that unit, placing that unit end to end (*iterating*) alongside the object' (Clements & Stephan, 2004, p. 300). Consider, for instance, a 30 cm ruler (a fairly standard measuring tool): the unit of measure is the centimetre, and the ruler as an object has been subdivided into thirty of these units. Each of these units is iterated—end to end—along the length of the ruler. We can then use this tool with its iterated units to make a measurement of a different object by placing these centimetre units along the object to be measured.

FIGURE 14.7 Unit iteration—'the teddies have to hold hands'

(Source: Amy MacDonald)

Taking a more informal approach, Figure 14.7 demonstrates the process of unit iteration using small plastic teddies as the unit of measure. To measure the paper, the children must iterate the unit (place the teddies side by side with no gaps, 'holding hands') along the length of the paper. Then, the children can count the number of iterated units, recognising that the last number counted represents the length measurement.

Typically, children first learn to measure using **informal units**—for example, pencils and straws for length, tiles or shapes for area, cups or small containers for volume, blocks or marbles for mass (McPhail, 2007), or indeed, the choice of unit shown in Figure 14.8—before progressing to the use of **formal units**, which are standardised units of measurement (for example, centimetres). As Bobis, Mulligan, and Lowrie (2013) explain, 'generally, the selection of a standard unit arises from the student's ability to measure more than one object with the same informal unit' (p. 156). Studies by Maranhãa and Campos (2000), and Stephan and Cobb (1998), have found that when children's measurement learning experiences include both informal and formal units, such as those shown in Figure 14.9, their understanding of measurement, units, and instruments is increased.

Informal units are non-standard units used in a consistent manner.

Formal units are standard units of measurement.

FIGURE 14.8 Using informal units to measure height

(Source: Paige Lee)

FIGURE 14.9 Exploring the relationship between informal and formal units

(Source: Amy MacDonald)

Pause and reflect

What strategies might you use to help children make connections between informal and formal units? Or how might you assist children to transition from using informal units to using formal units?

CURRICULUM CONNECTIONS

Australian Curriculum: Mathematics

Year one:

Measure and compare the lengths and capacities of pairs of objects using uniform informal units (ACMMG019)

- Understanding that in order to compare objects, the unit of measurement must be the same size.

Year two:

Compare and order several shapes and objects based on length, area, volume and capacity using appropriate uniform informal units (ACMMG037)

- Comparing lengths using finger length, hand span or a piece of string
- Comparing areas using the palm of the hand or a stone
- Comparing capacities using a range of containers.

(ACARA, 2022)

A key point to note is that units are arbitrary—they are used in different ways by different people in different contexts to suit different purposes. The big idea that we need to help develop in children is understanding that units—in many and varied forms—are a tool we can use to calculate measurements of objects, and communicating those measurements to others in meaningful ways.

Opinion Piece 14.1

DR SARAH FERGUSON

Recently I was supervising young children during free play time and noticed many of these children were spontaneously engaging in activities that involved measurement. Linking blocks in a long line that stretched across the room, putting items in a balance scale to try to make it 'balance', and pouring sand from one container to another were just some of the activities I noticed. I believe measuring is inherently engaging for children. Educators of young children can harness this curiosity by orchestrating rich measurement experiences that can lead to important mathematics learning.

I believe that measurement can be the vehicle for learning other concepts in mathematics rather than a separate unit of study usually done after more 'formal' mathematics is taught. Sometimes teaching counting or recording numbers can be lacking in purpose if we are just counting and writing numbers for no reason. Counting becomes important in measurement so we can answer 'how many blocks are as long as me'. Recording numbers will tell someone how many cups of sand this bucket holds. Measurement gives a rationale for both learning and practising these skills. There are also important concepts that children learn through measurement such as conservation, transitive reasoning and iteration. I think measurement is the key to both engaging young children's curiosity and teaching vital mathematical thinking.

Sarah Ferguson has taught in primary schools for more than twenty years mainly with children in their first or second year of school. She has been involved in research projects investigating children's learning of mass concepts, problem solving and inquiry, and multiplicative ideas. Sarah has taught pre-service teachers in mathematics education as well as in-service teachers in schools. She has an interest in children's drawings as a window into their learning and thinking and in the mathematical conversations that teachers can have with children.

Learn more
Go online to hear this Opinion Piece being read.

Chapter summary

In this chapter I have presented the measurement concepts that can be explored in early childhood mathematics education. Many children begin with emergent processes of measurement before developing more proficient processes that utilise formal units and tools. Measurement processes are underpinned by knowledge of measurable attributes, comparison, and units. Measurement also involves space and number concepts, as children make measurements of spatial objects and connect those measurements to number.

The specific topics we covered in this chapter were:

- emergent and proficient measurement
- measurable attributes
- length
- area
- volume and capacity
- mass and weight
- time
- temperature
- value (money)
- comparison
- units.

FOR FURTHER REFLECTION

1 What opportunities for measuring exist in young children's everyday lives?
2 How can everyday uses of measurement be embedded in the early childhood mathematics education program?
3 How can the language of measurement be modelled for young children?
4 In what ways might children show you their understanding of measurement concepts and processes?

FURTHER READING

Bobis, J., Mulligan, J., & Lowrie, T. (2013). Using measurement to make links. In *Mathematics for children: Challenging children to think mathematically* (4th ed., pp. 148–73). Frenchs Forest, NSW: Pearson Education Australia.

Jorgensen, R., & Dole, S. (2011). Measurement. In *Teaching mathematics in primary schools* (2nd ed., pp. 275–306). Crows Nest, NSW: Allen & Unwin.

Montague-Smith, A., & Price, A.J. (2012). Measures: Making comparisons. In *Mathematics in early years education* (3rd ed., pp. 145–73). New York: Routledge.

Stelzer, E. (2005). Measurement. In *Experiencing science and math in early childhood* (pp. 143–60). Toronto, ON: Pearson Education Canada.

15 Number and Algebra

Learn more
Go online to see a video from Paige Lee explaining the key messages of this chapter.

CHAPTER OVERVIEW

This chapter will explore the foundational number and algebra concepts that are developed in the early childhood years, present examples of these concepts, and provide an example Learning Experience Plan that shows how number and algebra can be explored with young children.

In this chapter, you will learn about the following topics:

- whole numbers
- subitising
- counting principles
- counting strategies
- number lines
- number operations
- additive and multiplicative thinking
- algebra
- pattern
- change
- function
- equivalence.

KEY TERMS

- Abstraction principle
- Addition
- Additive thinking
- Algebra
- Algebraic thinking
- Arithmetic
- Cardinal number
- Cardinal principle
- Change
- Conceptual subitising
- Counting back
- Counting on
- Derived facts
- Division
- Empty number line
- Equivalence
- Fact retrieval
- Function
- Functional thinking
- Multiplication
- Multiplicative thinking
- Nominal numbers
- Number
- Number line
- Number operations
- Number sense
- One-to-one principle
- Order irrelevance principle
- Ordinal number
- Pattern
- Perceptual subitising
- Rational counting
- Rote counting
- Semi-structured number line
- Skip counting
- Stable order principle
- Structured number line
- Subitising
- Subtraction
- Whole numbers

AMY MACDONALD

Introduction

Number
refers to values expressed by words or symbols used to represent a quantity of something.

Number sense
is about having understanding of, and fluency with, numbers and their relationships, size, and operations.

This chapter begins by introducing the foundational number ideas that allow us to perceive, and work with, quantities or collections of objects. **Number** refers to values expressed by words or symbols used to represent a quantity of something. In essence, number allows us to quantify our world. Number is often considered one of the more 'basic' aspects of mathematics; however, it is an incredibly complex and diverse construct encompassing a range of concepts, processes, principles, and operations. It is our understanding of, and fluency with, these concepts, processes, principles, and operations that endow the abstract idea of *number* with *meaning*. These attributes of understanding and fluency are typically referred to as **number sense**; that is, one's level of ease and familiarity with numbers (Jorgensen & Dole, 2011). Jorgensen and Dole explain this in greater detail:

> It is about understanding number meanings, knowing relationships between numbers, knowing the size of numbers, and knowing the effects of operating on numbers ... Number sense is never complete; developing number sense is a lifelong process that is promoted through many and varied experiences with using and applying numbers (pp. 130–1).

There is much research to suggest that babies are aware of quantity from birth. Antell and Keating (1983) found that the ability to represent the numerosity of small sets appears to exist in the first week of life. A significant body of research conducted in the 1980s and early 1990s found that infants are able to represent and remember quantities of up to three and sometimes four (Geary, 1994). Infants can also distinguish between quantities, match small sets of objects, and notice changes in quantities (Montague-Smith & Price, 2012). These skills become more advanced during the toddler years, and children soon learn to use counting words and connect these words to sets of objects. In doing so, children also begin to use the language of comparison as they distinguish between quantities (Montague-Smith & Price, 2012). By about four to five years of age, children have a good understanding of the connection between counting words and quantities, and they begin to extend these understandings to numbers greater than ten—and often well beyond!

It has been argued that the development of number understandings is the most important aspect of mathematics learning (Sarama & Clements, 2009). This, researchers suggest, is because early number understandings and number *sense* assist children in learning more complex mathematical ideas. However, it is often the case that meaningful understandings about number are developed through the exploration of other mathematical ideas such as measurement or shape; therefore, it is not necessarily the case that number should always be the first area of mathematics explored with young children. Indeed, the sorts of engagement that very young children have with space, shape, and comparison, for example, often provide meaningful opportunities for them to begin to develop number sense.

What do you recall about the ways in which number and algebra concepts and skills were introduced to you as a child?

CURRICULUM CONNECTIONS

Early Years Learning Framework

Outcome 5: Children are effective communicators

Children interact verbally and non-verbally with others for a range of purposes.

This is evident when children:

- Demonstrate an increasing understanding of measurement and number using vocabulary to describe size, length, volume, capacity and names of numbers
- Use language to communicate thinking about quantities to describe attributes of objects and collections, and to explain mathematical ideas.

Educators promote this learning when they:

- Listen to and respond to children's approximations of words
- Model language and encourage children to express themselves through language in a range of contexts and for a range of purposes
- Include real-life resources to promote children's use of mathematical language.

Children begin to understand how symbols and pattern systems work.

This is evident when children:

- Use symbols in play to represent and make meaning
- Begin to be aware of the relationships between oral, written and visual representations
- Draw on their experiences in constructing meaning using symbols.

Educators promote this learning when they:

- Engage children in discussions about symbol systems, for example, letters, numbers, time, money and musical notation
- Encourage children to develop their own symbol systems and provide them with opportunities to explore culturally constructed symbol systems.

(DET, 2019, pp. 43, 46)

Whole numbers

As outlined in the Introduction to this chapter, there are many aspects of number sense, but perhaps the basic element is the recognition of whole numbers. **Whole numbers** are positive numbers, including zero. They don't include fractions or decimals. Whole numbers and different representations of these numbers are all around us in everyday life, and children begin to perceive them from a very young age.

Whole numbers are positive numbers, including zero.

Knowledge of whole numbers typically begins by building up an awareness of numbers from one to ten. The possibilities for exploring these numbers are boundless: singing songs—think of 'Five Little Ducks' or 'Ten Cheeky Monkeys'—is a common method for building number awareness; another is counting fingers and toes. Children will often begin by recognising numbers of personal significance, such as their age or house number—and this early number recognition may well extend to numbers larger than ten, depending on the personal significance of the numbers.

Indeed, it is often the case that very young children delight in very large numbers, speculating about what numbers such as 'thousands' or 'millions' might mean in a given context, or even making up their own large numbers such as 'a million gazillion'.

All of these processes are important for developing an awareness of, and appreciation for, numbers and number systems, and these early applications and inventions of number form an important basis on which to build more formalised understandings.

CURRICULUM CONNECTIONS

Australian Curriculum: Mathematics

Year one:

Recognise, model, read, write and order numbers to at least 100 (ACMNA013)

- Modelling numbers with a range of materials and images.

Year two:

Recognise, model, represent and order numbers to at least 1000 (ACMNA027)

- Recognising there are different ways of representing numbers and identifying patterns going beyond 100
- Developing fluency with writing numbers in meaningful contexts.

(ACARA, 2022)

TYPES OF WHOLE NUMBERS

While we often talk about 'numbers' in a very general sense, it is important for children to learn that numbers can be used in different ways and for different purposes. Three important number uses to learn about are cardinal numbers, ordinal numbers, and the nominal use of numbers.

Cardinal number is a number used to label 'how many' in a set.

A **cardinal number** is a number that is used to label 'how many' in a set. This is sometimes referred to as recognising the 'how many-ness' of a number—such as the 'three-ness of three'—which basically means knowing that the word 'three' represents the *quantity* of 'three'. Figure 15.1 shows a common way of developing young children's understandings of cardinal numbers.

Ordinal number is a numerical representation used to explain the position of something in an order.

An **ordinal number** is a numerical representation used to explain the position of something in an order. We often use ordinal numbers as words to articulate a procession of discrete items, each distinguished by its position relative to the others—for example, 'first', 'second', 'third', and so forth. An ordinal number reminds us that numbers have a sequential relationship; that is, 'three' occupies a place in a sequence that is one more than 'two' and one fewer than 'four' (Montague-Smith & Price, 2012).

Nominal number is a number used as a name or label to help us identify something.

Finally, numbers are often used as **nominal numbers**; that is, as names or labels to help us identify things. However, it is important to note that there may not be a link between the nominal number name and its cardinality or ordinality—for example, just because a bus is called 'the Number 13 bus' doesn't necessarily mean that there is anything 'thirteen-ish' or 'thirteenth' about it (Montague-Smith & Price, 2012)!

FIGURE 15.1 Developing understandings of cardinal numbers

(Source: Michelle Muller)

> **CURRICULUM CONNECTIONS**
>
> *Australian Curriculum: Mathematics*
>
> Foundation year:
>
> *Compare, order and make correspondences between collections, initially to 20, and explain reasoning* (ACMNA289)
>
> - Understanding and using terms such as 'first' and 'second' to indicate ordinal position in a sequence
> - Using objects which are personally and culturally relevant to students.
>
> (ACARA, 2022)

Subitising

It is often the case that young children are encouraged to jump straight from numeral recognition to complex uses of counting. However, doing this excludes a very important precursor to counting—a skill known as 'subitising'. **Subitising** is the process of recognising how many items are in a small group (Knaus, 2013). There are two different forms of subitising: perceptual subitising and conceptual subitising.

Subitising is the process of recognising how many items are in a small group.

As the name implies, **perceptual subitising** refers to the instant recognition of the number of items in a small group. A child performs perceptual subitising without having to

Perceptual subitising refers to the instant recognition of the number of items in a small group.

count each item in the group or break the group into smaller parts. For this reason, perceptual subitising usually accommodates smaller collections, say, up to five. These collections are often displayed using fairly standard arrangements—those seen on dominoes, for example (Figure 15.2).

FIGURE 15.2 Subitising with dominoes

(Source: Amy MacDonald)

However, there are of course many ways in which perceptual subitising can be performed, and when practising this recognition skill it is very useful for young children to make use of diverse formats. For example, Figure 15.3 shows an alternative arrangement of three objects, while Figure 15.4 shows a columnar assemblage of five objects.

Conceptual subitising involves mentally breaking down larger collections and subitising the smaller groups.

A more complex version of subitising is known as **conceptual subitising**. This skill is used to subitise larger collections of objects. Conceptual subitising is a staged process. It involves the mental breaking of larger collections of objects into smaller, more familiar arrangements, recognising how many objects comprise the smaller groups, and then combining these quantities to determine how many objects make up the total collection. For example, when presented with a collection of ten items, a child using conceptual subitising might recognise two groups of five, and then combine these to make ten (Figure 15.5).

FIGURE 15.3 Another common layout for 'three'

FIGURE 15.4 A column approach is often used to represent collections

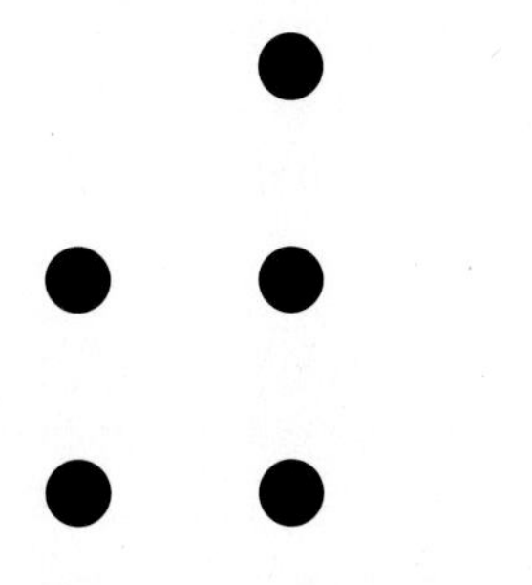

FIGURE 15.5 Conceptual subitising of 'ten'

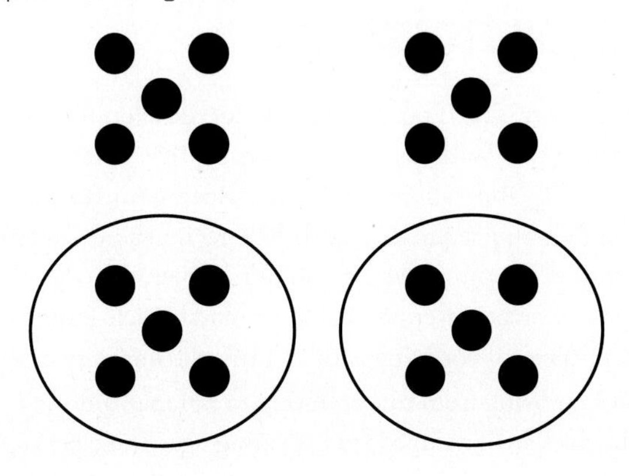

FIGURE 15.6 Perceptually subitising 'nine'

Of course, there will be many children who are quite capable of perceptually subitising larger numbers, without needing to break these into smaller groups. Often this is because familiarity with a particular arrangement of objects enables instant recognition of quantity. For example, early years educator Nikki Masters gives an example of a young child who was able to perceptually subitise the number nine in the arrangement shown in Figure 15.6, without having to 'chunk' into a group of four and a group of five. The reason the child gave for being able to do this was because he was familiar with this arrangement from watching television—this arrangement of nine dots is the logo for Australian television station Channel 9. This example highlights the importance of numbers in context!

CURRICULUM CONNECTIONS

Australian Curriculum: Mathematics

Foundation year:

Subitise small collections of objects (ACMNA003)

- Using subitising as the basis for ordering and comparing collections of numbers.

(ACARA, 2022)

Counting principles

Rote counting refers to memorising and reciting numeral names and sequences.

Rational counting refers to matching numeral names to groups.

Counting is a key number skill and is essential for developing competence with later arithmetic skills (Baroody & Wilkins, 1999; Knaus, 2013). There are two distinct forms of counting: rote counting and rational counting. **Rote counting** occurs when children memorise numeral names and sequences, and then recite these—often with the order of numbers slightly muddled, or numbers left out altogether. On the other hand, **rational counting** is demonstrated when a child is able to match each numeral name to a group, recognising that the purpose of counting is to identify the quantity of a group.

To assist children in moving from rote counting to rational counting, it is important that the five counting principles are developed from a young age. These principles are as follows:

1. one-to-one principle
2. stable order principle
3. cardinal principle
4. order irrelevance principle
5. abstraction principle.

These important principles will now be explored in greater depth.

ONE-TO-ONE PRINCIPLE

One-to-one principle refers to the need to match one counting word to each item in the set to be counted.

The **one-to-one principle**, or one-to-one correspondence, as it is more commonly known, refers to the need to match one and only one counting word to each item in the set to be counted (Montague-Smith & Price, 2012). This process involves dividing the set of items to be counted into two groups: the group for which each item has already been given a counting word, and a group for which each item is yet to be given one. Montague-Smith and Price (2012) describe the skills that children need to develop in order to demonstrate the one-to-one principle. First, children need to be able to recite the counting words in order. Then, they need to coordinate touching the objects to be counted and saying the counting words so that these happen at the same time. Touching (or pointing to) the objects is an important aspect of this process as it ensures that each object is counted only once. This also helps children to keep track of which objects have and have not been counted—another key skill. The process is also aided by placing the objects to be counted in a straight line. When developing the one-to-one principle, children usually find it helpful to move the objects as they count them. Accordingly, it is generally better to start by having children count real objects rather than objects represented in pictures, as this allows the children to move the objects around as they see fit.

STABLE ORDER PRINCIPLE

The **stable order principle** refers to the realisation that the counting sequence stays consistent, and that number names are used in a fixed order every time (Knaus, 2013). This is initially developed through rote counting, where children memorise and chant the counting words in order, without necessarily attaching meaning to these words. However, over time these words and the order in which they are said takes on meaning and becomes related to the items to be counted (Montague-Smith & Price, 2012). In particular, when learning to count beyond twenty, children must learn to recognise the pattern within each decade (i.e. 21, 22, 23 ... 31, 32, 33 ... 41, 42, 43 ... and so on), as well as in the naming of each new decade (i.e. 20, 30, 40 ... and so on).

Stable order principle refers to the realisation that the counting sequence stays consistent.

CARDINAL PRINCIPLE

The **cardinal principle** refers to an understanding that the last counting word said when counting a set of objects represents the total number of objects in the set. So, for example, when a child counts 'one-two-three-four-five teddies', knowledge of the cardinal principle means that the child will know that there are five teddies in the set. Knowledge of both the one-to-one principle and the stable order principle is essential for developing the cardinal principle.

Cardinal principle refers to an understanding that the last counting word said represents the total number of items.

ORDER IRRELEVANCE PRINCIPLE

The **order irrelevance principle** refers to an understanding that when counting objects, you can begin with any object in the group and the total will remain the same. This principle is related to the cardinal principle in that it involves recognising that cardinality is not affected by the order in which the objects are counted—for example, there are still five teddies regardless of whether you start by counting the first teddy, the middle teddy, or the last teddy.

Order irrelevance principle refers to the fact that when counting, it doesn't matter which object is counted first the total will remain the same.

ABSTRACTION PRINCIPLE

The **abstraction principle** refers to the knowledge that when objects are counted, the number of objects in the group is the same regardless of whether the group is made up of similar or different objects (Knaus, 2013). For example, there might be five teddies in a group, or there might be two teddies, one dinosaur, one robot, and one Matchbox car—regardless, there are still five objects in the group. This can be quite a tricky concept for children to learn, as it requires them to recognise that all the other counting principles apply, regardless of what the objects themselves actually are.

Abstraction principle refers to the knowledge that when objects are counted, the number of objects in the group is the same regardless of whether they are similar or different in nature.

A summary of the counting principles is presented in Table 15.1.

TABLE 15.1 Counting principles

One-to-one principle	One counting word for each item in the set to be counted
Stable order principle	Counting sequence stays consistent
Cardinal principle	Last counting word said represents the total number of items
Order irrelevance principle	Regardless of which item in a set is counted first, the total remains the same
Abstraction principle	Number of items in the set is the same regardless of whether they are similar or different items

CURRICULUM CONNECTIONS

Australian Curriculum: Mathematics

Foundation year:

Connect number names, numerals and quantities, including zero, initially up to 10 and then beyond (ACMNA002)

- Understanding that each object must be counted only once, that the arrangement of objects does not affect how many there are, and that the last number counted answers the 'how many' question
- Using scenarios to help students recognise that other cultures count in a variety of ways, such as the Wotjoballum number systems.

(ACARA, 2022)

Learning Experience Plan 15.1

'Counting with magnets'

Experience information

This learning experience is planned for five children aged between 3 years and 3-and-a-half years who attend the centre together. These children always play together and are part of a little Einstein's program in our room designed to continue to engage and further stimulate the older children. The children will be using paddle-pop sticks numbered from one to ten, with a magnetic strip on each, to count and sort paper clips to the corresponding numeral. The children will also be learning about magnetic forces.

Learning focus

I have chosen to plan this learning experience as over the past two weeks I have observed the children in a number of play experiences that included different number concepts. The children were using these concepts in almost all play experiences. As numbers have become an interest for the group of children, I plan to incorporate counting and sorting into a game in which the children will extend on their knowledge of numbers but will also be learning about magnets and magnetic forces.

This experience will enhance the children's mathematical learning as the children will be counting, matching the correct number of paper clips to each paddle-pop stick, and sorting the paddle-pop sticks into the correct numerical order. Through Bishop's mathematical activities of *counting*, *locating*, and *explaining*, the children will be able to count how many paddle-pop sticks there are. These activities will be extended through the use of one-to-one correspondence, rational counting, and the cardinal principle, which have already been observed in the children's play. The paddle-pop sticks will be numbered from one to ten, which will allow the children to sort the paper clips into number groups and stick the designated number of paper clips onto the correspondingly numbered paddle-pop stick using ordering and number recognition. The children will then be able to explain exactly how many paper clips are on each paddle-pop stick as they express their ideas of number with the visual representation of numerals in front of them.

EYLF outcomes

This learning experience will create an active learning environment to support the children in developing dispositions for learning such as curiosity, enthusiasm, and cooperation. The children will be working towards several of the EYLF outcomes during this experience. Outcomes 4 and 5 are most relevant as the aim is for the children to become confident and involved learners as well as effective communicators in terms of their learning. As the experience will occur in a small group, the children will be able to contribute to discussions effectively and use reflective thinking.

Another outcome that the children will work towards is to transfer what they learnt from one context, which would be their own play experiences, to another context, which is the experience that is planned for them. Throughout the experience the children will be interacting both verbally and non-verbally to share their ideas and contribute to the groups learning. An increasing understanding of number through the use of vocabulary will also be developed. These outcomes will be facilitated as I will model mathematical language, scaffold their learning, and encourage the children to investigate and solve problems.

Requirements

This experience will be set up on a table on the deck outside the room to allow the children to be free from distractions and to help them concentrate. A positive and supportive social and physical environment will be created through the use of a small group and positive interactions to help the children thrive during the activity. I will be sitting at the table with the children to help guide and scaffold the experience. Before the children come outside, I will set up the experience with all the necessary materials.

The materials needed for this learning experience are:

- a table
- six chairs
- decking
- ten thick paddle-pop sticks with magnetic strips already glued onto each one
- fifty-five metal paper clips
- a plastic zip-lock bag
- a small container for paper clips.

Procedure

- Once the experience is set up on the deck, I will invite the small group of children outside for the activity.
- Children will be invited to join me at the table. When the children are sitting at the table, I will first talk to them about what we are going to do and what we will be learning (i.e. magnets, numbers, sorting paper clips).
- I will then take the paddle-pop sticks out of the bag and lay them on the table, asking the children if they know what numbers are on the sticks. I will ask the following questions. 'What numbers can you see on the paddle-pop sticks?' and 'Which number is this?'
- I will then encourage the children to count the numbers on each paddle-pop stick, from one to ten, and ask them if they can put the sticks in the correct order.
- Once the paddle-pop sticks are in a correct order, I will encourage the group to count from one to ten all together as I point to the numbers.

- To engage the children and to develop their curiosity, I will ask them what they think is on the paddle-pop sticks (the magnets).
- I will then take out the container of paper clips and ask the children what they think will happen when they put a paper clip near one of the magnetic strips.
- After listening to all of their predictions, I will then ask the children one by one to put a paper clip near a magnet. 'What happened?' and 'What is the paper clip doing?' are two questions I will ask the children.
- Once everyone has had a turn, I will then ask the children, 'Why do you think this has happened?' and 'Do you know why the paper clip got stuck to the paddle-pop stick?' and listen to their responses.
- I will then talk to the children about magnets and how it was the magnetic fields that pulled the paper clip towards the magnet through attraction. At this stage, I will also explain to the children that not all objects are attracted to magnets. I will then ask the children to find different objects on the deck to test whether or not they will stick to the magnet.
- The container of paper clips will be placed on the table, and I will go around one by one asking the children to place the same number of paper clips on the paddle-pop stick as denoted by the numeral that appears on the stick. 'How many paper clips did you place on the paddle-pop stick?'—I will ask this question when they have finished. This will promote the further development of one-to-one correspondence, knowledge of cardinal numbers, and rational counting. The children will be able to have two turns at matching the paper clips to the numerals.
- When all the children have had a turn, we will count the paper clips on each paddle-pop stick from one to ten to finish the experience.
- I will ask the children to help me collect all the paper clips and place them back into the container.
- The children will then be transitioned into indoor play.
- Once all children are inside, I will pack away the materials so they can be used again.

Plan for review

During the experience, observations will be written in the form of notes and photographs will be taken as well. If notes haven't been taken during the experience, then observations will be completed as soon as it is finished to ensure that an accurate and critical evaluation can take place. It is important to note any information that the children expressed, things that worked well as well as things that didn't, so that the experience can be evaluated in suitable detail. The way the children are engaged and how much they are engaged, the way they interact and how they explore the experience are all things that should be observed. It is important to reflect not only on the experience, but also on the children's learning and the educator's personal observations. Could things have been done differently to enhance the experience?

(Source: Samira Smith)

Counting strategies

When counting, children may use a range of different strategies. Effective use of counting strategies is a precursor to the development of the four number operations (i.e. addition, subtraction, multiplication, division) (Jorgensen & Dole, 2011). The three main strategies that children use are counting on, counting back, and skip counting.

COUNTING ON

Counting on occurs when children begin with a given number and count on from this (Jorgensen & Dole, 2011). For example, if a child is asked to count on from the number 4, they would be expected to continue the counting sequence from the number 4 (i.e. 5, 6, 7, 8, and so on). It is important that children develop the ability to *begin* counting from the given number, rather than revert to starting at the number 1. Concrete materials such as number lines, calendars, and hundreds charts are all useful ways of supporting children to count on from a given number.

Counting on occurs when children begin with a given number and count on from this.

The ability to count on is essential for developing addition skills. Children can solve the addition of two numbers by counting on from the first number by the amount of the second number. For example, 6 + 4 can be solved by starting at the number 6 and then counting on four more (i.e. 7, 8, 9, 10 to reach the solution that 6 + 4 = 10).

COUNTING BACK

Counting back is the reverse of counting on, and it requires children to identify 'what comes before' any given number (Jorgensen & Dole, 2011). For example, if a child is asked to count back from the number 4, they would be expected to continue the reverse counting sequence from the number 4 (i.e. 3, 2, 1). Counting back assists children to develop subtraction skills. When subtracting one number from another, children can count back the smaller number from the larger number. For example, 6 – 4 can be solved by starting at the number 6 and then counting back four numbers (i.e. 5, 4, 3, 2 to reach the solution the 6 – 4 = 2).

Counting back requires children to identify what comes before any given number.

SKIP COUNTING

Skip counting is counting in multiples of any number other than one (Jorgensen & Dole, 2011). Children often begin to practise skip counting by counting up by twos (i.e. 2, 4, 6, 8, and so on). Other common multiples that are introduced early on are tens (i.e. 10, 20, 30, 40, and so on) and fives (i.e. 5, 10, 15, 20, and so on). Skip counting encourages children to recognise patterns in number sequences, which is an important precursor to multiplication and division (Jorgensen & Dole, 2011).

Skip counting is counting in multiples of any number other than one.

CURRICULUM CONNECTIONS

Australian Curriculum: Mathematics

Foundation year:

Establish understanding of the language and processes of counting by naming numbers in sequences, initially to and from 20, moving from any starting point (ACMNA001)

- Reading stories from other cultures featuring counting in sequence to assist students to recognise ways of counting in local languages and across cultures
- Identifying the number words in sequence, backwards and forwards, and reasoning with the number sequences, establishing the language on which subsequent counting experiences can be built

- Developing fluency with forwards and backwards counting in meaningful contexts, including stories and rhymes
- Understanding that numbers are said in a particular order and there are patterns in the way we say them.

Year one:

Develop confidence with number sequences to and from 100 by ones from any starting point. Skip count by twos, fives and tens starting from zero (ACMNA012)

- Using the popular Korean counting game (sam-yuk-gu) for skip counting
- Developing fluency with forwards and backwards counting in meaningful contexts such as circle games.

Year two:

Investigate number sequences, initially those increasing and decreasing by twos, threes, fives and tens from any starting point, then moving to other sequences (ACMNA026)

- Developing fluency and confidence with numbers and calculations by saying number sequences
- Recognising patterns in number sequences, such as adding 10 always results in the same final digit

(ACARA, 2022)

Number lines

A number line is a line on which numbers are marked at intervals.

Structured number lines have all the marks for numbers.

Semi-structured number lines have some of the marks for numbers.

Empty number lines have no markings.

A **number line** is a line on which numbers are marked at intervals and is commonly used to represent numerical operations such as addition and subtraction. **Structured number lines** (Figure 15.7) have all the marks for the numbers; **semi-structured number lines** (Figure 15.8) have some of the marks for numbers; and **empty number lines** (Figure 15.9) have none of the markings and allow children to draw marks for themselves (Siemon et al., 2011).

FIGURE 15.7 Structured number line

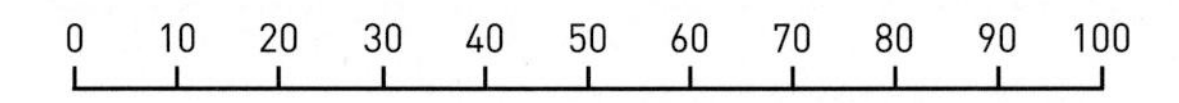

FIGURE 15.8 Semi-structured number line

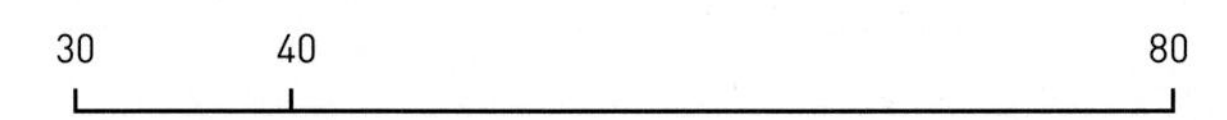

FIGURE 15.9 Empty number line

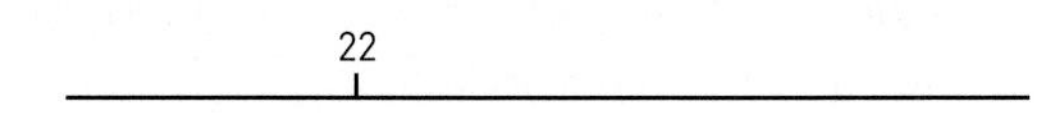

Number lines can be used to model and represent a range of mathematical ideas and processes. For example, a structured number line can be used to remind children that numbers have a sequential relationship; that 'three' is one more than 'two' and one fewer than 'four' (Montague-Smith & Price, 2012). Number lines can also be used to model counting—in particular, skip counting. Skip counting provides a useful means of modelling 'jumps' on a number line, which can be a precursor to, and useful for, number operations.

CURRICULUM CONNECTIONS

Australian Curriculum: Mathematics

Year one:

Recognise, model, read, write and order numbers to at least 100. Locate these numbers on a number line (ACMNA013)

- Identifying numbers that are represented on a number line and placing numbers on a prepared number line.

Year two:

Describe patterns with numbers and identify missing elements (ACMNA035)

- Describing a pattern created by skip counting and representing the pattern on a number line.

(ACARA, 2022)

Number operations

Number operations refer to the processes through which we perform calculations with numbers (often referred to as **arithmetic**). The four number operations are addition, subtraction, multiplication, and division. Research indicates that many children can use their informal mathematical knowledge to carry out simple number operations before receiving any formal instruction in relation to these (Baroody & Wilkins, 1999). In particular, situations involving sharing help young children to develop understandings of part/whole, more/less, adding/taking away, and fairness/equal distribution. These informal understandings are critical for the development of more formal addition, subtraction, multiplication, and division operations.

Number operations refer to the processes of addition, subtraction, multiplication, and division.

Arithmetic involves manipulating, and calculating with, numbers.

ADDITION

Put simply, **addition** refers to adding parts to make a whole. Of course, a range of synonyms might be used in place of adding: joining, pairing, meeting, putting altogether are but a few examples. The simplest of addition problem types are those that involve combining two sets. These are called *aggregation* problems, meaning 'bringing together' or 'combining'. A more challenging type of addition problem concerns growth, known as an *augmentation* problem, or 'increasing' problem (Montague-Smith & Price, 2012).

Addition refers to adding parts to make a whole.

Geary (1994) identified five general strategic classes that young children use for solving simple addition problems. These are:

1 using manipulatives
2 finger counting

3 verbal counting without the use of manipulatives (i.e. mentally)
4 derived facts
5 fact retrieval.

The use of manipulatives, or concrete materials, allows children to model an addition problem in a tangible way. Finger counting is another form of concrete modelling that allows children to visualise the addition problem as they work through it. Both methods also allow children to 'keep track' of their counting as they solve the addition.

As previously described, children can use various counting strategies to assist them in solving addition problems. Counting on, in particular, is a form of verbal counting that allows children to solve addition problems, often without the need for manipulatives.

Derived facts are memorised addition facts used to solve addition problems.

Derived facts are memorised addition facts used as a strategic basis for solving addition problems (Geary, 1994). Young children typically memorise 'doubles' such as 1 + 1 = 2, 2 + 2 = 4, 3 + 3 = 6. Other common addition facts are structured around the Base-10 system (i.e. 'making to ten'). Ten-frames—two-by-five rectangular frames into which counters or other objects can be placed—are useful tools for exploring addition facts for numbers less than, or equal to, ten (Figure 15.10).

FIGURE 15.10 Using ten-frames to explore addition facts

(Source: Amy MacDonald)

Fact retrieval refers to the use of memorised answers to solve addition problems.

Finally, many children are able to use **fact retrieval** to solve simple addition—that is, they 'just know it', quickly producing the answer without overt signs of counting (Geary, 1994). Children come to memorise answers through their experiences with counting and derived-facts strategies. For instance, children may be able quickly to identify that 1 + 2 = 3 because they are aware of the fact that 3 follows 1 and 2 in the counting sequence (Siegler & Shrager, 1984).

CURRICULUM CONNECTIONS

Australian Curriculum: Mathematics

Foundation year:

Represent practical situations to model addition and sharing (ACMNA004)

- Using a range of practical strategies for adding small groups of numbers, such as visual displays or concrete materials
- Using Aboriginal and Torres Strait Islander methods of adding, including spatial patterns and reasoning.

(ACARA, 2022)

SUBTRACTION

Building on from addition, **subtraction** refers to the removal or withdrawal of one part from a whole to find the remaining part. A range of everyday words are used to represent this process—for example, take away, less, remove, and so forth. Like addition, subtraction has a range of structures and problem types. The two most commonly explored in the early years are *take away* problems, where one or more items are taken away from a set; and *difference* problems, which are used to compare two numbers or sets of objects (Montague-Smith & Price, 2012).

Subtraction refers to removing a part from a whole to find the remaining part.

Children typically utilise the same five strategies described for addition when solving subtraction problems. Counting-based strategies—either counting on or counting back, and with or without manipulatives or fingers—are the most commonly used subtraction strategies for young children (Geary, 1994).

CURRICULUM CONNECTIONS

Australian Curriculum: Mathematics

Year one:

Represent and solve simple addition and subtraction problems using a range of strategies including counting on, partitioning and rearranging parts (ACMNA015)

- Developing a range of mental strategies for addition and subtraction problems

Year two:

Solve simple addition and subtraction problems using a range of efficient mental and written strategies (ACMNA030)

- Becoming fluent with a range of mental strategies for addition and subtraction problems, such as commutativity for addition, building to 10, doubles, 10 facts and adding 10
- Modelling and representing simple additive situations using materials such as 10 frames, 20 frames and empty number lines.

(ACARA, 2022)

MULTIPLICATION

Multiplication refers to grouping the parts within a whole.

Multiplication is typically associated with 'grouping' and is often framed in terms of how many 'lots of' a certain number there are. However, children work with other forms of multiplication problems beyond grouping. These include *rate* problems (i.e. comparing one quantity to a different quantity), *scalar* problems (i.e. identifying 'how many more times than'), and *cross-product* problems (i.e. identifying different combinations). Exploration of these types of problem typically occurs in the later primary years, but the foundations for the capacity to deal with such problems are laid from the very early explorations of multiplication.

CURRICULUM CONNECTIONS

Australian Curriculum: Mathematics

Year two:

Recognise and represent multiplication as repeated addition, groups and arrays (ACMNA031)

- Representing array problems with available materials and explaining reasoning
- Visualising a group of objects as a unit and using this to calculate the number of objects in several identical groups.

(ACARA, 2022)

DIVISION

Division refers to dividing a whole into parts.

Division is concerned with dividing a whole into parts. A range of words are used to describe this process, such as sharing, splitting, and partitioning. There are two basic forms of division problem: *partition* problems and *quotition* problems. The difference between the two is related to the way in which the whole is shared. In partition division, the number of groups is known, but how many will be in each group is unknown. On the other hand, in quotition problems, the amount in each group is known, but the number of groups is unknown (Jorgensen & Dole, 2011). It is important for children to experience both types of division problem.

Fractions

Children's early experiences with fair sharing and partitioning play a crucial role in the development of a sense of fractions (Siemon et al., 2011). A *fraction* is a ratio of parts to a whole; as such, children's early experiences of sharing and splitting into parts form a conceptual foundation for later explorations of fractions. Young children understand what it means to cut objects into smaller, but roughly equal, parts; for example, cutting an apple in half, cutting an orange into quarters, or cutting a pizza into eighths (Siemon et al., 2011). Through such experiences, children learn the language of fractions (for example, 'halves', 'quarters'), and develop an awareness of the division processes that result in such fractions.

CURRICULUM CONNECTIONS

Australian Curriculum: Mathematics

Year one:

Recognise and describe one-half as one of two equal parts of a whole (ACMNA016)

- Sharing a collection of readily available materials into two equal portions
- Splitting an object into two equal pieces and describing how the pieces are equal.

Year two:

Recognise and represent division as grouping into equal sets and solve simple problems using these representations (ACMNA032)

- Dividing the class or a collection of objects into equal-sized groups
- Identifying the difference between dividing a set of objects into three equal groups and dividing the same set of objects into groups of three.

Recognise and interpret common uses of halves, quarters and eighths of shapes and collections (ACMNA033)

- Recognising that sets of objects can be partitioned in different ways to demonstrate fractions
- Relating the number of parts to the size of a fraction.

(ACARA, 2022)

Additive and multiplicative thinking

While the four operations of addition, subtraction, multiplication, and division are often treated separately, it is perhaps more beneficial for children to explore the operations in tandem, learning that when particular operations are paired, they 'undo' one another. Specifically, addition and subtraction can be seen as pairs in that on the addition side, parts join to give wholes, while on the subtraction side, knowing the whole and one part allows the other part to be found (Booker, Bond, Sparrow, & Swan, 2014). Understanding of these relationships between addition and subtraction is referred to as **additive thinking**. Similarly, multiplication and division are connected in that multiplication involves grouping parts to form a whole, while division involves separating a whole into parts. This is known as **multiplicative thinking**. Multiplicative thinking is essential for the development of more complex mathematics, as it underpins concepts such as fractions, ratio, and proportion, which are developed later in the primary school years.

Additive thinking refers to understanding of the operations of addition and subtraction, and of the relationships between the two.

Multiplicative thinking refers to understanding of the operations of multiplication and division, and of the relationships between the two.

CURRICULUM CONNECTIONS

Australian curriculum: mathematics

Year two:

Explore the connection between addition and subtraction (ACMNA029)

- Becoming fluent with partitioning numbers to understand the connection between addition and subtraction
- Using counting on to identify the missing element in an additive problem.

(ACARA, 2022)

Opinion Piece 15.1

Learn more
Go online to hear this Opinion Piece being read.

DR ANN DOWNTON

When it comes to thinking about number and algebra, I think of a broader picture of mathematics being a study of patterns and relationships and how these permeate much of what young children do when they are playing games, doing puzzles or exploring their environment. Observing young children engage with these everyday activities made me appreciate the rich foundation for number learning that they provide. Underpinning these everyday experiences are mathematics concepts such as conservation of number, more and less, cardinality, counting, equality, sequencing, and pattern recognition. When exploring these and other key ideas such as subitising and combining and partitioning numbers as teaching educators, it is important to utilise these familiar experiences and to assist young children to see the relevance of their learning and to make connections. A simple exploration with dominoes can lead to children recognising pattern in the dot arrangements, as well as noticing doubles, sorting and classifying, and quantifying. As teacher educators, it is about exploring these ideas in a holistic way rather that in bite-size pieces. Related to this is the importance of utilising resources such as having large floor tens frames where children can stand in a square to act out a story of children being on a bus or boat, and if others join imagine 'how many now' and 'how they know'. Experiences such as this engage children in learning about quantities, as well as early addition and subtraction and recognising odd and even numbers.

My recent research with four- and five-year-old children into early ideas of multiplication and division using games, real-life context and children's literature highlighted that young children are capable of more than we realise. I experienced this when I observed five-year-old children playing a game of 'go go group'. Children move to music and when the music stops they get into groups of a particular size. One boy said that 3 would not work because there were only 20 children present this day. He recognised that 20 was not divisible by 3 and proved it by counting by 3s. Similarly, when researching young children's engagement with challenging tasks, some children when asked to work out how many spider legs were on a large family of spiders in the quickest way, some grouped the legs into twos and counted them, others worked out the number of legs on one spider then doubled and doubled or counted by fours. Such a task allows the children to think flexibly about the counting unit they might choose to count, look for patterns and notice structural relationships. As teacher educators, we need to provide learning experiences that build on and foster their curiosity to want to explore further.

Ann Downton is a Senior Lecturer in undergraduate and post-graduate early childhood and primary mathematics education courses in the Faculty of Education at Monash University, Clayton, Australia. She has extensive experience as a primary school teacher, a part-time mathematics specialist and mathematics consultant. Ann's research interests include children's mathematics learning, particularly their development of multiplicative thinking, pedagogical practices that foster children's mathematical thinking, and formative assessment practices. The focus of her recent research study was on challenging sequences of learning that enhance the cognitive and affective experiences of Foundation to Year 2 students when learning mathematics. Ann's current research focuses on young children's intuitive ideas of multiplication and division prior to formal instruction.

Additive and multiplicative thinking both underpin the development of early algebraic and functional thinking. Algebra, as it relates to early childhood mathematics classrooms, will be explored in the next section of this chapter.

Algebra

Algebra can be defined as generalised arithmetic, where children take their early learning about number and move to seeing generalisations (Jorgensen & Dole, 2011). As Hunter and Miller (2022) explain, 'the ability to generalise mathematical structures, by noticing a common structure that can be applied to multiple cases, is fundamental to mathematics' (Kaput, 1999; Radford, 2010; Warren & Cooper, 2008). The ability to identify and generalise mathematical structures is critical for later learning in mathematics and STEM (science, technology, engineering, and mathematics) more broadly, as it underpins skills such as problem solving and computational thinking (Miller, 2019).

Algebra
refers to generalised arithmetic.

The basis for algebra learning is established in the early years, when children begin to notice regularities in their calculations with numbers (MacGregor & Stacey, 1999). Young children are capable of generalising mathematical structures across a range of contexts, including generalising relationships between numbers and pattern rules in real-life situations (e.g. number of puppy dog ears in relation to dog tails) (Hunter & Miller, 2022). Algebra focuses on *properties* of operations, and on *relations* among operations, rather than simply on *carrying out* operations, which is the main focus in arithmetic (Jorgensen & Dole, 2011). This way of thinking about operations can be described as algebraic thinking. **Algebraic thinking** is about generalising the structure of arithmetic. According to Siemon et al. (2011, p. 256), early algebraic thinking entails:

Algebraic thinking
is about generalising the structure of arithmetic.

- making explicit the mathematics of pattern (especially repeating and growing patterns), and extending these patterns to our number system
- studying early functional thinking with a focus on the relationships between the operations, such as the inverse relationship between addition and subtraction
- studying the structure of the number system and operations—for example, the meaning of equals and the meaningful use of unknowns. It does not entail the use or manipulation of symbols.

When do you remember first learning about algebra, explicitly?

PATTERN

In Chapter 12 we explored the key concepts associated with patterns, with particular focus on the ways in which patterns can be created, copied, and continued. A pattern is a sequence of two or more items that repeat. When children begin to describe, generalise, and justify these patterns, they are developing skills that are critical for algebraic thinking. These processes help children to understand the structure of arithmetic, which forms the basis for algebraic understanding (Siemon et al., 2011). Research has shown that early development of one's ability to recognise pattern and structure enables a stronger foundation for algebraic thinking and has a positive influence on mathematical achievement overall (Mulligan, Mitchelmore, & Prescott, 2006).

CHANGE

Change refers to a process through which something becomes different.

Exploring the concept of change provides another opportunity for children to think algebraically (Copley, 2010). **Change** refers to a process through which something becomes different. As the National Council of Teachers of Mathematics (NCTM) (2000) explains:

> The understanding that most things change over time, that many such changes can be described mathematically, and that many changes are predictable helps lay a foundation for applying mathematics to other fields and for understanding the world (p. 95).

In the early years, children can explore different patterns to identify consistent changes. Children can also explore changes that come about through reversal, such as the inverse relationship between addition and subtraction, and multiplication and division (Siemon et al. 2011). Exploring changes is a useful way of introducing the idea of function to young children (Copley, 2010).

FUNCTION

A function is a rule that describes a pattern.

Functional thinking refers to understanding of the relationship between two or more varying quantities.

Functions involve consistent change (Siemon et al., 2011). In basic terms, a **function** is the rule that describes a pattern. **Functional thinking** focuses on the relationship between two (or more) varying quantities (Siemon et al., 2011, p. 256). 'Function machines' or 'Guess the rule' games are a fun way of exploring functions with young children. In these types of games, children need to focus on the change from the input to the output in order to understand the input–output relationship (Reys, Lindquist, Lambdin, & Smith, 2012). Hilton, Dole, and Campbell (2014) provide a simple explanation of how function machines work:

> [The educator] records particular numbers that 'go in' to the function machine and the number that 'comes out'. Children suggest input numbers until a pattern is established. They can then try to state the rule that describes the pattern (i.e. the function) that has been applied to each incoming number to produce the corresponding outgoing number (p. 164).

Function machines can be created in lots of different ways:

- A function machine can simply be drawn on the whiteboard or a piece of cardboard—draw a box with a funnel going in and a funnel coming out (you might like to give the machine a face to add some character).

- There a lots of function machine templates available on the web—try doing a Google Images search for 'function machines'.
- There are also interactive function machines available on the web. These are great if you have a smartboard or data projector in your room. A Google web search for 'function machines' will give you lots of different ones to try.
- You might like to make an actual function machine with the children—what a great way to use up some boxes, tubes, and whatever other craft resources you have lying around!

EQUIVALENCE

Equivalence refers to the state of being equal.

Equivalence, or the state of things being equal, is a key concept of algebraic thinking. The equals sign needs to be carefully introduced so that children understand that the meaning of the sign is that the value on one side of the sign is equal to the value on the other side. Number sentences are often articulated as 'making' an answer, such as '2 plus 3 makes 5'. However, the use of the word 'makes' does not indicate the notion of equivalence, and this can result in the alternative conception that the equals sign simply means 'the answer is coming' (Siemon et al., 2011). This alternative conception can be a barrier to algebraic thinking (Jorgensen & Dole, 2011). To avoid such a scenario, educators should use the language of equivalence when exploring number sentences with children, such as '2 plus 3 is equal to 5'. To reinforce the notion of equivalence, children can model number sentences with concrete materials such as counters or blocks. Balance scales or number beam balances are also a useful means of modelling equivalence (see Figure 15.11).

FIGURE 15.11 Modelling equivalence with a number beam balance

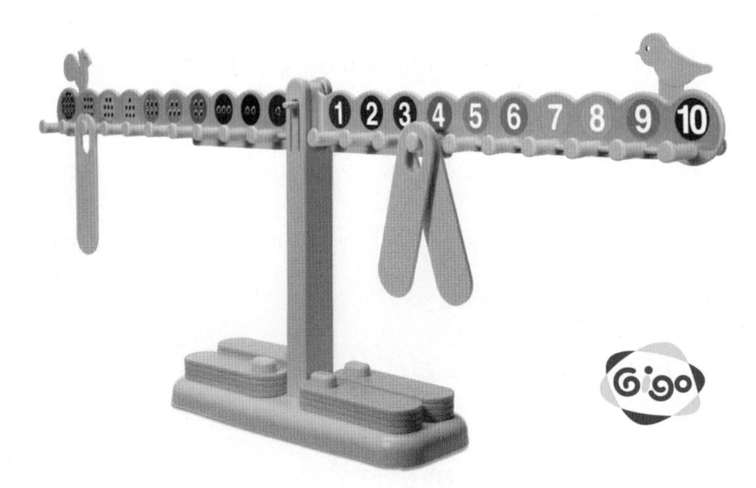

Pause and reflect

In what other ways might the concept of equivalence be modelled for young children?

Chapter summary

In this chapter I have presented the foundational number and algebra concepts that can be explored in early childhood mathematics education. Facility with number concepts and skills forms the basis for number sense, and an ability to generalise the structure of arithmetic is the basis for algebraic thinking. Both number sense and algebraic thinking are critical for later mathematics learning in the schooling years.

The specific topics we covered in this chapter were:

- whole numbers
- subitising
- counting principles
- counting strategies
- number lines
- number operations
- additive and multiplicative thinking
- algebra
- pattern
- change
- function
- equivalence.

FOR FURTHER REFLECTION

To help you plan for number and algebra learning experiences, think about the following:

1. In what ways do you use number and algebra understandings to perform everyday tasks?
2. What are some examples of young children's everyday explorations of number and algebra?
3. What are some concrete resources that might be used to assist children in developing number and algebra concepts?
4. How might you model appropriate number and algebra language with young children?

FURTHER READING

Booker, G., Bond, D., Sparrow, L., & Swan, P. (2014). Computation: Additive thinking. In *Teaching primary mathematics* (5th ed., pp. 195–265). Frenchs Forest, NSW: Pearson Australia.

Booker, G., Bond, D., Sparrow, L., & Swan, P. (2014). Computation: Multiplicative thinking. In *Teaching primary mathematics* (5th ed., pp. 266–360). Frenchs Forest, NSW: Pearson Australia.

Copley, J.V. (2010). Patterns, functions, and algebra in the early childhood curriculum. In *The young child and mathematics* (2nd ed., pp. 79–98). Reston, VA: National Council of Teachers of Mathematics.

Jorgensen, R., & Dole, S. (2011). Early number. In *Teaching mathematics in primary schools* (2nd ed., pp. 130–60). Crows Nest, NSW: Allen & Unwin.

Knaus, M. (2013). Early number experiences. In *Maths is all around you: Developing mathematical concepts in the early years* (pp. 33–48). Albert Park, VIC: Teaching Solutions.

Montague-Smith, A., & Price, A.J. (2012). Calculating and problem solving with number. In *Mathematics in early years education* (3rd ed., pp. 52–82). New York: Routledge.

Montague-Smith, A., & Price, A.J. (2012). Number and counting. In *Mathematics in early years education* (3rd ed., pp. 23–51). New York: Routledge.

Reys, R., Lidquist, M.M., Lambdin, D.V., & Smith, N.L. (2012). Algebraic thinking. In *Helping children learn mathematics* (10th ed., pp. 300–21). Hoboken, NJ: John Wiley & Sons, Inc.

16 Data, Statistics, and Probability

Learn more
Go online to see a video from Paige Lee explaining the key messages of this chapter.

CHAPTER OVERVIEW

This chapter will explore the foundational data, statistics, and probability concepts that are developed in the early childhood years, present examples of these concepts, and provide an example Learning Experience Plan that shows how data, statistics, and probability can be explored with young children.

In this chapter, you will learn about the following topics:

» data
» sorting and grouping
» statistics
» collecting and organising data
» representing data
» interpreting and communicating data
» probability.

KEY TERMS

» Chance
» Data
» Probability
» Statistical literacy
» Statistics

Introduction

This chapter outlines the big ideas associated with data, statistics, and probability. Young children are continually posing questions that need to be solved, and one way of answering these questions is through data collection (Knaus, 2013).

The foundational skills required for the collection of data include sorting and matching. The handling of data requires an ability to sort information into meaningful groups. Research has shown that the skills of sorting and matching develop in infancy. By about twelve months, babies are able to choose objects belonging to the same group, and by about eighteen months, children are able to form groups using items of their own choosing (Montague-Smith & Price, 2012). These skills become progressively more sophisticated as children move from sorting by one attribute, to sorting by more than one attribute, and ultimately, they learn to state the rule by which a collection has been sorted—even if it has been sorted by someone else. These skills are usually evident from around four to six years of age (Montague-Smith & Price, 2012).

CURRICULUM CONNECTIONS

Early Years Learning Framework

Outcome 5: Children are effective communicators

Children begin to understand how symbols and pattern systems work.

This is evident when children:

- Begin to sort, categorise, order and compare collections and events and attributes of objects and materials, in their social and natural worlds
- Draw on memory of a sequence to complete a task.

Educators promote this learning when they:

- Provide children with access to a wide range of everyday materials that they can use to create patterns and to sort, categorise, order and compare.

(DET, 2019, pp. 43, 46)

Learning about data in the early childhood years is an important precursor to learning about statistics in the years to come. Statistics is based on processes of collecting, recording, describing, displaying, and organising data (Booker, Bond, Sparrow & Swan, 2014). These skills are essential for using information to make decisions, judgements, and interpretations.

The ability to make critical judgments about data is known as **statistical literacy**. It is this skill that allows us to discern whether the narrative being presented about a given set of data is accurate and fair. It is important to recognise that data can be manipulated and used in ways that misrepresent the information gathered; as such, statistical literacy is a critical mathematical skill to be developed.

Statistical literacy is the ability to make critical judgments about the ways in which data are presented and used.

Statistical literacy also requires an understanding of the chance or likelihood of something happening. We encounter risk and uncertainty in all aspects of our life, so developing an understanding of probability assists children to make decisions that are associated with risk (Siemon et al., 2011). In short, understandings of data, statistics, and probability provide essential life skills for making judgments about everyday situations.

Data

Data
is an area of mathematics in which information is collected and analysed in order to find out about our world.

Data essentially means information collected and analysed in order to find out about our world. Data is an area of mathematics that can at times be overlooked; however, it is often something that children and their educators gain a great deal of enjoyment out of investigating together (Bennett & Weidner, 2014). One of the main reasons for this is that data is an area of mathematics that is easily applied to everyday life and provides children with endless opportunities for exploring topics that are interesting to them (Bennett & Weidner, 2014). From birth, children will have been making sense of their worlds by categorising and identifying attributes of objects through everyday activities at home, such as sorting the washing or putting their toys away (Montague-Smith & Price, 2012). These early experiences with sorting and grouping lay important foundations for developing more sophisticated understandings about handling data.

Sorting and grouping

In order to sort and group objects, children need to have a concept of the attribute they are using as the basis for comparison (Montague-Smith & Price, 2012). Even very young children are able to identify attributes of objects and compare objects on the basis of the similarity of the particular attributes. For example, young children recognise the difference between things (for example, different colours, different shapes), and know that some things go together (for example, knife and fork, socks and shoes) (Knaus, 2013). Playing games and puzzles that involve matching (see Figure 16.1) are a simple way of helping children to develop concepts of 'same' and 'different'.

One of the most everyday ways in which children learn about sorting and grouping objects is through packing away their play things. As shown in Figure 16.2, the act of stacking items away on the shelves in an orderly manner might involve creating groups of similar items and arranging these items on the shelves accordingly.

FIGURE 16.1 Matching puzzle

(Source: Penny Baker/Amy MacDonald)

FIGURE 16.2 Sorting items when packing them away

(Source: Kathryn Hopps)

What other aspects of the early childhood program provide opportunities for children to sort and classify? Consider, for example, the potential of meal times, or transitions between activities.

Pause and reflect

CURRICULUM CONNECTIONS

Australian Curriculum: Mathematics

Foundation year:

Sort and classify familiar objects and explain the basis for these classifications (ACMNA005).

(ACARA, 2022)

Learning Experience Plan 16.1

'Sorting rubbish and creating a display for our bins'

Experience information

Planned for the following children:

E, 4 years 11 months

J, 4 years 2 months

A, 5 years

K, 4 years 5 months.

Learning focus

This learning experience is designed to develop the children's process of exploring by problem solving how to sort materials into containers as well as assisting with the problem of mixed rubbish bins. I hope that this problem solving will involve questioning and further researching using the interactive whiteboard to connect with technology. I wish also to extend the children's concept of explaining as they express the factual information that they may obtain from research.

The learning experience is intended to develop the processes of identifying numbers and counting as the children classify and sort the materials. Here we could also explore and identify quantities.

Also to be explored in the learning experience will be the concept of selecting and using appropriate resources while discussing the use of equipment/tools as the children create signs for the bins.

Finally, the children will engage in a play-based learning experience, which should develop the mathematical processes involved in playing. A play-based approach such as this will encourage the children to interact with the key points of mathematics as they sort the materials and explore quantity and representation.

EYLF outcomes

During this experience, the children will be working towards developing a range of the learning outcomes from the *Early Years Learning Framework*. I have designed this experience with a learning focus aligned predominantly to Outcome 4: 'Children are confident and involved learners'. The experience is framed to extend on the learning and development of the children and is based on developmental areas that have become evident from my observations. These areas of development include:

- developing dispositions for learning, in particular curiosity, cooperation with others, creativity, confidence, and enthusiasm
- developing skills of problem solving, researching, and furthering inquiries with use of technology.
- resourcing learning while connecting with people and technologies.

The experience will also give the children opportunity to extend on development as determined in Outcome 2, as they learn to become socially responsible for their environment, and in Outcome 5, as they use information technologies to investigate ideas.

Requirements

I will be setting up this experience in the indoor environment to allow the children to sit together on the mat with access to all of the resources. This will allow me to target the specific age group and their needs/interests. The mat is situated in front of the interactive whiteboard. I'll need to move the experience to a table while the children create their bin signs.

Materials needed:

- assortment of clean rubbish, food, and recyclable materials
- three containers for sorting (green, red, and yellow) to match the bin colour coding
- interactive white board
- creative expression materials such as pencils, textas, and paper.

I will present the experience to the small group of children by setting up the coloured containers on the mat with the container of mixed rubbish materials in front. I will have the interactive whiteboard behind the mat turned on and ready for use. I will set up a nearby table with pencils, textas, and paper to allow extension of the experience to creative expression. I will invite the children inside and revisit their recent learning/interests by talking to them about what I have observed.

Procedure

I hope to engage the children's interest in this educator-instigated experience by involving them in reflective discussion to revisit their recent interests/learning. I will talk to the children about what I have observed and invite them to extend on this through an experience that I need their help to undertake. I hope that the children will be immediately interested in partaking in the experience. The experience that I have planned should extend on their interests and involve them in

- use of the interactive whiteboard
- helping with tasks, including sorting and classifying
- producing creative expression/writing
- engaging with different resources
- participating in small-group peer interactions.

I will ask, 'Would you like to help me with a sorting experience with some rubbish? I have noticed lots of rubbish going into the wrong bins and I would really appreciate your help.' I hope this type of question will ignite the children's interest in the experience. I am hoping to maintain their interest in the experience by asking them to help to sort the materials into the correct containers. I will be talking to the children about how I have observed the waste being put into the wrong bins and asking for their assistance so that the waste is correctly sorted and allocated.

During the experience, I will use the strategy of encouraging feedback to assist the children with understanding their learning and make them feel recognised for their participation. I will use the strategy of indirect teaching by leading the children's learning and asking appropriate questions to encourage their thinking and inquiry skills. I may also have to use direct teaching if the children need guidance with sorting the specific materials into the correct containers or to guide their learning with technology. This direct teaching might need to be used while the children are creating their signs for the bins, but I will try not to be too directive—instead, I will ask questions to encourage the children to make their own discoveries. I plan to be involved in the experience to guide their learning, but in a way that facilitates the development of their own understandings and extends on their investigation skills. I will ask such questions as, 'Where can we go to find out what coloured bin this goes in?' This will hopefully engage the children in furthering the mathematical, scientific, and technological learning that is the purpose of the experience. I will use questions that ask the children to complete, correct, delete and sort, compare and organise, to prompt their thinking and inquiry throughout the experience.

I have planned this experience to engage the children in sorting and classifying while extending on inquiry skills. I am hoping that it will progress to making signs for our bins to assist with the problem of incorrect waste disposal. I think that this experience has the potential to be extended on to build the children's interests and development in the future, and it may become an ongoing project. I don't feel that I will need to end the experience; instead, I will let it evolve to follow the small group's interests/learning.

After the children have completed their signs, I will encourage them to be respectful to their learning environment by tidying up the resources. I plan to involve the children in the clean-up of the rubbish materials by taking the filled containers and emptying them in the bins. This could also provide opportunities for further learning as we take these bins out to the skip bins at the end of the day.

Plan for review

After the experience, I plan to review and reflect on it to conclude if it was successful. I will determine if the learning focus was appropriate by observing the children's development and assessing if they were capable of attempting the tasks while further extending on their development. I will know if my strategies were effective in encouraging the learning focuses by the level of engagement from the children and their initiative to further their own inquiries. If the experience is too teacher led, I will not gain any new understandings of the children's learning/development. Through observing the children and analysing the experience, I will be able to determine whether the children's development was facilitated and whether their learning has progressed. If the children show little or no engagement, I will know that the experience was not suitable. Using the *Early Years Learning Framework's* process of an ongoing cycle of planning, documenting, and assessing children's learning as well as reflective practice, I will be able to think about any improvements that could have been made or to determine if the learning areas were successfully developed. Through this reflective practice I will self-reflect on my teaching practices to ensure that my role in the experience was appropriate and facilitated the children's learning.

I will be observing how the children engage with this experience to help me to evaluate my teaching and the desired learning outcomes. Listening to the children's questions and their responses, observing their body language, and watching their small-group interactions will all assist me with reflecting on the experience and further planning.

(Source: Michelle Call)

Statistics

Statistics refers to the processes of collecting, recording, describing, displaying, and organising data.

An ability to sort on the basis of identified characteristics is prerequisite to developing the data-handling skills required to create, use, and understand statistics. **Statistics** can be defined as the processes of collecting, recording, describing, displaying, and organising data (Booker et al., 2014). This section explores the different ways in which children learn to handle data and generate statistics. This whole process is sometimes known as the PCRAI cycle (Montague-Smith & Price, 2012, p. 176), which stands for:

- Pose the question related to the problem identified
- Collect the data
- Represent the data
- Analyse the data
- Interpret the data in terms of the original question, and if there is still a problem a new question may need to be formulated and the cycle repeated.

However, there are many variations to the PCRAI cycle, so this chapter explores the skills involved in data handling in more general terms.

COLLECTING AND ORGANISING DATA

The first step in collecting data is to pose a question for investigation. It is important that children have the opportunity to pose questions that are of interest to them—that is, what would *they* like to find out? We can assist this process by modelling questions such as, 'I wonder what our favourite books are?' or 'How might we find out what type of food we should have at our class party?'

Once the question for investigation has been posed, children need to discuss the different strategies that might be used to help them answer the question. The key thing for children to understand is that they need to be able to gather information that can be counted or measured in some way. Strategies might include asking people to draw a picture or answer a question; alternatively, information may be gathered simply by observing what happens at particular times (for example, conducting a traffic survey by counting cars as they pass is a classic data-gathering activity). Children also need to talk about strategies for keeping track of the data they have gathered—that is, how will they record the information they accumulate? For this reason, collecting pictures (or something similar) is a useful first step as it gives the children tangible data to 'collect', literally. From there, children can be extended to use recording techniques such as coloured dots, stickers, or tally marks.

The next step is for the children to organise their data into meaningful groups and work out how many responses they have in each group. This task requires children to have understandings of concepts such as *alike*, *same*, *different*, *belongs*, or *doesn't belong* (Knaus, 2013). For example, if children were conducting a survey as to the class's favourite party food, they might need to group different kinds of lollies into one group known collectively as 'lollies', different types of chips (such as potato and corn) into a group called 'chips', and so forth. Sorting and organising data in this way is an essential step to be completed before meaningful representations of the data can be created.

What opportunities for data collection exist in the early childhood education setting? How might the data collected by the children be used to inform the educational program?

CURRICULUM CONNECTIONS

Australian Curriculum: Mathematics

Foundation year:

Compare, order and make correspondences between collections, initially to 20, and explain reasoning (ACMNA289)

- Comparing and ordering items of like and unlike characteristics using the words 'more', 'less', 'same as' and 'not the same as' and giving reasons for these answers.

Answer yes/no questions to collect information and make simple inferences (ACMSP011)

- Posing questions about themselves and familiar objects and events
- Representing responses to questions using simple displays, including grouping students according to their answers
- Using data displays to answer simple questions such as 'how many students answered 'yes' to having brown hair?'

Year one:

Choose simple questions and gather responses and make simple inferences (ACMSP262)

- Determining which questions will gather appropriate responses for a simple investigation.

Year two:

Identify a question of interest based on one categorical variable. Gather data relevant to the question (ACMSP048)

- Determining the variety of birdlife in the playground and using a prepared table to record observations.

Collect, check and classify data (ACMSP049)

- Recognising the usefulness of tally marks
- Identifying categories of data and using them to sort data.

(ACARA, 2022)

REPRESENTING DATA

Once the data has been collected and organised into meaningful groups, it is important for children to learn about the different ways in which those data might be represented. In this section we will look at some simple forms of representing data that might be explored with young children.

Picture graphs

Picture graphs are a very concrete way of representing data that has been collected. There is no real need for labels at this point; however, once children have become familiar with organising data in this format, the educator can gradually introduce labels such as those shown in Figure 16.3 (i.e., 'number of children', 'types of icecream') (Jorgensen & Dole, 2011).

FIGURE 16.3 Picture graph

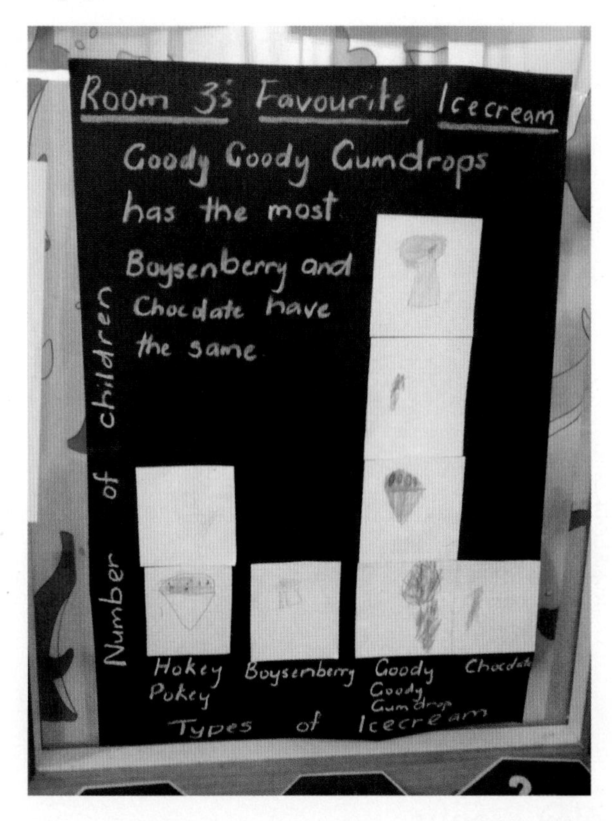

(Source: Maree Parkes/Amy MacDonald)

Simple picture graphs of this kind emphasise the need for a common baseline along which the pictures are placed, thus allowing for accurate comparison of the groups of pictures. This is a stepping stone into the construction of more formal column graphs.

CURRICULUM CONNECTIONS

Australian Curriculum: Mathematics

Year one:

Represent data with objects and drawings where one object or drawing represents one data value. Describe the displays (ACMSP263)

- Understanding one-to-one correspondence
- Describing displays by identifying categories with the greatest or least number of objects.

Year two:

Create displays of data using lists, table and picture graphs and interpret them (ACMSP050)

- Creating picture graphs to represent data using one-to-one correspondence
- Comparing the usefulness of different data displays.

(ACARA, 2022)

Column graphs

The drawing of column graphs requires an extension of the labelling of axes to include simple scales. The scales assist children in transferring the data collected into a graphical form. They also assist the reader of the graph in making quick interpretations of the data, rather than having to count, individually, the items represented, as is necessary in a picture graph. Figure 16.4 shows how the data displayed in the picture graph in Figure 16.3 might be represented as a column graph.

FIGURE 16.4 Column graph

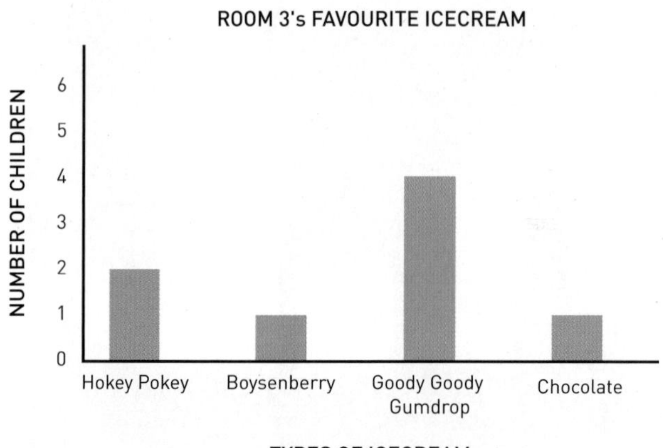

More complex graphs

As children progress through their schooling years, they will be introduced to more complex ways of graphing data. Common forms that they will learn about include *comparative bar graphs* (Figure 16.5), which are used to compare two sets of data; *grouped frequency histograms* (Figure 16.6), which show clusters of continuous data; and *circle or pie graphs* (Figure 16.7), perhaps the most challenging graph of all to construct. While these forms of graphs are not necessarily relevant to children in the early childhood years, it is helpful for us as educators to have sense of the learning continuum to which we are contributing when we explore simple picture or column graphs with children.

INTERPRETING AND COMMUNICATING DATA

The interpretation and communication of data are achieved by analysing and summarising the data to draw conclusions (Knaus, 2013). For example, in Figure 16.3 we can see that the group has reached the conclusion that 'Goody Goody Gumdrops' is the favourite icecream because it has the most pictures. They have also observed that 'Boysenberry' and 'Chocolate' have the same number of pictures.

FIGURE 16.5 Comparative bar graph

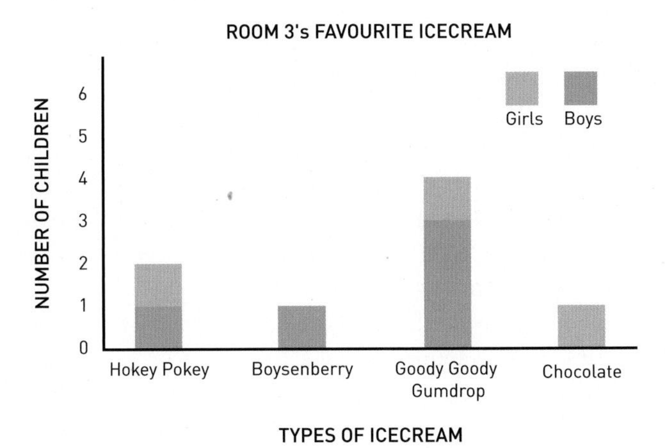

FIGURE 16.6 Grouped frequency histogram

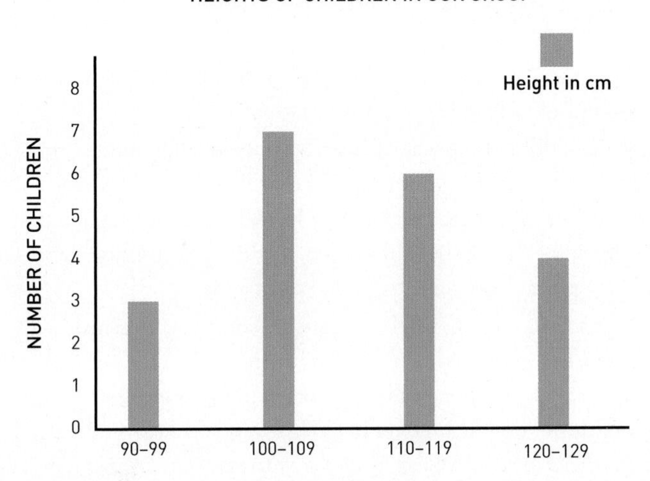

FIGURE 16.7 Circle ('pie') graph

According to Jorgensen and Dole (2011), while considerable time is spent teaching children how to *construct* graphs, less time is spent on teaching them how to *interpret* graphs. This is problematic because in today's society we must have the capacity to read and interpret data, and we must understand how data are being represented (or indeed, misrepresented). The critical reading of data is a key element of statistical literacy. Statistical literacy draws upon all of the big ideas discussed in this chapter: it involves being able to identify the types of data that need to be collected; collecting data and organising those data in meaningful ways; representing data in ways that make sense and have high levels of readability; and being able to interpret and critique data (Jorgensen & Dole, 2011). Furthermore, statistical literacy involves being able to interpret and evaluate information that is being presented by others and make judgments as to whether or not it is a fair and reasonable representation of those data.

Opinion Piece 16.1

Learn more
Go online to hear this Opinion Piece being read.

DR JAMES RUSSO

I believe that it is vital that young children are given opportunities to engage with, and explore, ideas around probability, data and statistics.

Probability is a difficult concept for both children and adults to grasp. Exposing young children to these ideas informally, both through games (for example, 'Is it possible to roll a 7 on this dice? How do you know?') and experiences (for example, Is it likely or unlikely that it will be rain today?), begins to familiarise them with relevant language and supports their capacity to think critically.

Data and statistics at all levels is fundamentally concerned with collecting, organising and interpreting information to help us answer questions and inform decision making. Such statistical literacy is particularly important this century, given the amount of information we all have to navigate and critically evaluate. The best way of introducing these ideas to young children is to tap into their innate curiosity and support them to become 'little scientists'. This involves:

helping them to pose questions that they are interested in (for example, what is the most popular football team in our kinder?)

encouraging them to make predictions (for example, Richmond Tigers)

deciding what information they need to locate or collect to answer their question (for example, ask everybody including the educators their favourite team and everybody is only allowed to give one answer)

helping them to find ways of communicating ('showing' and 'telling') their answers to others (for example, a pictograph).

To ensure the process is both purposeful and promotes a critical orientation, children should then be encouraged to reflect on their predictions ('Were they correct?

What was surprising'), and to ask follow-up questions ('Now you know this, what else do you want to know?'). This practice of 'thinking like a scientist' can be dialled up and down in terms of its sophistication depending on the age and developmental level of a child and can be modelled authentically by educators as they navigate 'adult problems' (for example, deciding where to display our class's latest art work).

James Russo is a Senior Lecturer at Monash University interested in working with educators of children to make mathematics more stimulating and enjoyable to teach and learn.

Probability

Generally, the topic of probability is about the likelihood or chance that a particular outcome will occur (Siemon et al., 2011). More formally, **probability** is the term used to describe a ratio expressing the chance of the occurrence of a certain event (Macmillan, 2009); for instance, 'there is a 1 in 13 983 816 chance of winning the lottery'. A probability may also be expressed as a percentage—for example, 'there is a 75% chance of rain tomorrow'.

Probability
is a ratio expressing the chance of the occurrence of an event.

In explorations of probability, it is useful to focus on the concept of chance. **Chance** is essentially about the certainty, possibility, or impossibility of the occurrence of an event (Macmillan, 2009). Siemon et al. (2011) suggest that a useful starting point is for children to classify events as likely or unlikely, or as impossible, possible, or certain. They give the following examples for consideration:

Chance
is about the certainty, possibility, or impossibility of the occurrence of an event.

- My aunty will visit on the weekend.
- My cat will talk.
- It will rain this afternoon.
- Things will fall if you drop them.
- My bedroom will tidy itself.
- Everyone in our class owns a dog.
- The sun will rise tomorrow.
- My mum is older than I am.
- My football team will win (p. 471).

Such prompts are likely to respond in interesting discussions among children. Children will respond differently to these prompts based on their own experiences with, and knowledge of, the topic. Differing perspectives should be encouraged, and children should be given the opportunity to articulate their reasoning for the likelihood they attribute to each prompt.

In what other ways might the concept of probability be modelled for young children?

CURRICULUM CONNECTIONS

Australian Curriculum: Mathematics

Year one:

Identify outcomes of familiar events involving chance and describe them using everyday language such as 'will happen', 'won't happen' or 'might happen' (ACMSP024)

- Justifying that some events are certain or impossible.

Year two:

Identify practical activities and everyday events that involve chance. Describe outcomes as 'likely' or 'unlikely' and identify some events as 'certain' or 'impossible' (ACMSP047)

- Classifying a list of everyday events according to how likely they are to happen, using the language of chance, and explaining reasoning.

(ACARA, 2022)

Chapter summary

In this chapter I have presented the foundational data, statistics, and probability concepts that can be explored in early childhood mathematics education. Knowledge of data, statistics, and probability is essential for everyday decision making, such as making judgments about risks and the likelihood of events. Young children can be supported to collect and analyse data in order to find out about their world.

The specific topics we covered in this chapter were:

- data
- sorting and grouping
- statistics
- collecting and organising data
- representing data
- interpreting and communicating data
- probability.

FOR FURTHER REFLECTION

Next time you are watching the news, note the instances in which data are presented or statistics are cited, and think about the following:

1. For what purposes have the data or statistics been used?
2. What influence do these data and statistics have on the way you perceive or interpret the story?
3. Is enough information provided to allow the viewer to judge the accuracy or relevance of the data presented?
4. How might these statistical literacy skills be introduced to young children?

FURTHER READING

Booker, G., Bond, D., Sparrow, L., & Swan, P. (2014). Statistics and probability. In *Teaching primary mathematics* (5th ed., pp. 509–41). Frenchs Forest, NSW: Pearson Australia.

Jorgensen, R., & Dole, S. (2011). Chance and data. In *Teaching mathematics in primary schools* (2nd ed., pp. 307–38). Crows Nest, NSW: Allen & Unwin.

Knaus, M. (2013). Probability and statistics (chance and data). In *Maths is all around you: Developing mathematical concepts in the early years* (pp. 77–88). Albert Park, VIC: Teaching Solutions.

Montague-Smith, A., & Price, A.J. (2012). Sorting, matching and handling data. In *Mathematics in early years education* (3rd ed., pp. 174–96). New York: Routledge.

Glossary

Abstraction principle
When objects are counted, the number of objects in the group is the same regardless of whether they are similar or different in nature.

Accumulation of distance
Knowing that as you iterate the unit and count the units, the numbers represent the space covered by the units.

Addition
Adding parts to make a whole.

Additive thinking
Understanding of the operations of addition and subtraction, and of the relationships between the two.

Affective domain
Beliefs, attitudes, and emotions about mathematics.

Algebra
Generalised arithmetic.

Algebraic thinking
Generalising the structure of arithmetic.

Area
The amount of space contained within a 2D shape.

Area array
A structure formed when a unit is iterated in two directions.

Arithmetic
Manipulating, and calculating with, numbers.

Array
An arrangement of rows and columns with equal numbers in each row and equal numbers in each column.

Aspirations
Positive hopes, such as those for children as they start school.

Assessment
The process of collecting, organising, and analysing information about children's performance.

Attitudes (mathematical)
Learnt responses towards mathematics.

Beliefs (mathematical)
Psychologically held understandings about mathematics.

Bi-directional influences
The relationships between the child and the settings in which the child participates.

Bishop's Mathematical Activities
Processes of counting, measuring, locating, designing, playing, and explaining.

Capacity
The amount an object can hold.

Cardinal number
A number used to label 'how many' in a set.

Cardinal principle
Realisation that the last counting word said represents the total number of items.

Centration
The tendency to fixate on a single attribute.

Chance
The certainty, possibility, or impossibility of the occurrence of an event.

Change
A process through which something becomes different.

Child-instigated experiences
Incidental learning experiences led by the child's curiosity.

Chronosystem
The socio-historical timeframe of the child's life.

Classification
The process of grouping objects according to specific attributes.

Cognition
Acquisition of knowledge through mental processes such as reasoning, planning, problem solving, representing, and remembering.

Column approach
A tool for implementing the strengths approach—provides a structured framework for thinking through the first five stages of the strengths approach, and applying them in practical ways.

Comparison
Making a judgment about the similarities or differences between two or more objects or events.

Competency cycle
A focus on strengths and positive expectations that results in the development of new competencies.

Complex circumstances
Social, financial, cultural, or health challenges experienced by families.

Concepts
The building blocks of knowledge that allow people to organise and categorise information.

Conceptual subitising
Mentally breaking down larger collections and subitising the smaller groups.

Concrete operational period
The organisation of concrete operations.

Congruent transformations
Alterations to the position of a shape.

Conservation
Understanding that attributes such as weight, length, and number remain constant despite changes in appearance.

Conservation of length
Knowing that if an object is moved, its length does not change.

Constructionism
A theory based on the belief that learners generate new ideas when they are engaged in constructing an artefact.

Constructivism
A theory based on the belief that learners construct their own knowledge.

Content strands
Curriculum strands that specify the mathematical topics to be developed.

Contexts
Interactions, activities, and settings.

Continuity
A sense of connection from one context to the next.

Counting
The process of expressing numerical quantifiers and qualifiers.

Counting back
Identifying what comes before any given number.

Counting on
Beginning with a given number and counting on from this.

Culturally responsive teaching practices
Practices that recognise and respond to the diverse cultural and language backgrounds of children and families.

Culture
The common ways of knowing and being that people in a community share.

Curriculum
Everything that happens through the day in an educational program.

Curriculum frameworks
Documents that guide practice and provide long-term outcomes for children's learning.

Data
An area of mathematics in which information is collected and analysed in order to find out about our world—also refers to the information so collected.

Decompose
Break a pattern down to its individual parts or deconstruct an object.

Deficit cycle
A focus on negative expectations that result in negative experiences.

Depth
How deep something is, along a single plane or through two dimensions.

Derived facts
Memorised addition facts used to solve addition problems.

Designing
The process of expressing a symbolic plan, structure, or shape.

Developmental theory
A theory based on notions or conjectures that focus on sequential, predictable stages of development.

Digital play
Play that incorporates digital technology.

Digital technology
Electronic tools, systems, devices, and resources.

Direct teaching
Offering conceptual cues and suggesting more effective strategies.

Dispositions
Characteristics that encourage children to respond in particular ways to learning opportunities.

Division
Dividing or splitting a whole into parts.

Documentation panels
Displays of work samples, recorded conversations, photographs, and anecdotal evidence.

Duration
How long an event takes.

Early childhood
The period from birth to eight years of age.

Early childhood mathematics education
Opportunities for learning about mathematics across the range of contexts in which young children participate.

Ecological theory
A theory based on the belief that children's development is influenced by the social and cultural systems within the child's environment.

Edge
Where two faces of a shape meet.

Educator-instigated experiences
Learning experiences that are planned in advance and are directed by the educator.

Emergent measurement
Children's meaningful use of measurement for their own purposes.

Emotions (mathematical)
Affective responses towards mathematics.

Empty number line
A number line with no markings for numbers.

Enactive representations
Use of real materials to depict relationships.

Enculturation
A process by which people learn the values and behaviour appropriate to a culture.

Entitlements
Rights, such as those that children have as they start school.

Equivalence
The state of being equal.

Evaluator role
Analysing, interpreting, assessing, and providing feedback.

Exosystem
The social system one step removed from the child.

Expectations
Ideas, beliefs, assumptions—such as those about what school will be like, and what it means to be a school student.

Explaining
The process of expressing factual or logical information.

External representations
Physical, observable configurations.

Face
The side of a solid shape.

Fact retrieval
The use of memorised answers to solve addition problems.

Family gatherings
Workshops designed to assist educators and families to explore mathematics together.

Fluency
Carrying out procedures flexibly, accurately, efficiently, and appropriately.

Formal mathematics
Formal and standardised practices and approaches to mathematics.

Formal operational period
Hypothesis making and testing the possible.

Formal units
Standardised units of measurement.

Function
A rule that describes a pattern.

Functional thinking
Understanding of the relationship between two or more varying quantities.

Geometric reasoning
The invention and use of formal systems to investigate shape and space.

Geometry
A knowledge domain that encompasses size, shape, relative position, and movement of two-dimensional figures in the plane and three-dimensional objects in space.

Growing patterns
Patterns that have a similar relationship between elements, but in which each element increases or decreases.

Growth mindset
The belief that intelligence can be learnt.

Height
How high something is, along a single plane or through two dimensions.

Iconic representations
Use of drawings or pictures to depict relationships.

Indirect teaching
Guiding play and investigation through thoughtful questioning and listening.

Informal mathematics
The use of informal mathematical practices in everyday life.

Informal units
Non-standardised units used in a consistent manner.

Instructor role
Providing and managing knowledge-acquisition processes.

Intentional teaching
Being active and purposeful in selecting pedagogies.

Internal representations
Mental configurations that are not directly observable.

Irreversible thinking
The inability to begin at the end and work backwards.

Learning Experience Plan
A structured planning document that articulates a learning focus, procedure, and review strategy for a learning experience.

Learning stories
Qualitative snapshots of learning recorded as written narratives, often with accompanying photographs.

Length
How long something is, along a single plane or through two dimensions.

Line of symmetry
A point of a 2D shape at which one side of the shape reflects the other.

Locating
The process of positioning oneself and other objects in space.

Logico-mathematical knowledge
Knowledge developed in the mind by thinking about an object.

Loose parts
Materials that can be moved, combined, taken apart, and put together in multiple ways.

Macrosystem
The wider cultural, social, and political context.

Mass
The amount of matter in an object.

Mathematical disposition
The inclination or tendency to use mathematics.

Mathematical identity
How people label and understand themselves in relation to mathematics.

Mathematical mindset
The belief in our ability to learn and improve in mathematics.

Mathematics
A knowledge domain that is concerned with patterns, relationships, representations, symbols, abstraction, and generalisations.

Mathematics anxiety
Feelings of tension and fear towards mathematical situations.

Measurement
The assignment of a numerical value to an attribute of an object or event.

Measuring
The process of ascertaining or calculating quantities or entities that cannot be counted or located spatially.

Mediator role
Shared meaning-making and facilitation of learning.

Mesosystem
Relationships between microsystems.

Microsystem
The immediate environments in which the child participates.

Modelling
Providing opportunities to model or exemplify, test, and check mathematical knowledge.

Multiplication
Grouping the parts within a whole.

Multiplicative thinking
Understanding of the operations of multiplication and division, and of the relationships between the two.

Nominal number
A number used as a name or label to help us identify something.

Noticing
A collection of practices for living in, and learning from, experience.

Number
A value expressed by a word or symbol used to represent a quantity of something.

Number line
A line on which numbers are marked at intervals.

Number operations
The processes of addition, subtraction, multiplication, and division.

Number sense
Having understanding of, and fluency with, numbers and their relationships, size, and operations.

Numeracy
A social and cultural perspective for discovering, thinking about, and applying mathematical knowledge.

One-to-one principle
The matching of one counting word to each item in a set to be counted.

Opportunities
The chances and possibilities available in a setting, such as those available to children as they start school.

Order irrelevance principle
Realisation that when counting, it doesn't matter which object is counted first—the total will remain the same regardless.

Ordinal number
A numerical representation used to explain the position of something in an order.

Outcomes
Skills, knowledge, and dispositions that educators can promote.

Partitioning
Dividing an object into equal-sized units.

Pattern
A sequence of two or more items that repeat.

Pedagogies of educational transitions
Approaches that help to establish connections across educational settings.

Perceptual dominance
The tendency to focus on visually striking features.

Perceptual subitising
Instant recognition of the number of items in a small group.

Physical knowledge
Information about the qualities or attributes of objects and what they are used for.

Planning
The process of using information to inform the provision of learning experiences.

Playful pedagogies
Approaches that are playful for all involved (adults and children).

Playing
The process of imitating or recreating social, concrete, or abstract models of reality.

Polygon
Any 2D shape with straight sides.

Preoperational period
Symbolic representation of the present and real world.

Probability
A ratio expressing the chance of the occurrence of an event.

Problem
A question that engages someone in searching for a solution.

Problem solving
Formulating, investigating, and resolving problems and communicating solutions.

Processes
The actions through which concepts are explored.

Proficiency strands
Curriculum strands that describe how the content is to be explored or developed.

Proficient measurement
Comprehension of measurement concepts, operations, and relations.

Projective transformations
Alterations involving enlargements or reductions.

Rational counting
Matching numeral names to groups.

Reasoning
Capacity for logical thought and actions.

Reflective practice
Examining what happens in educational settings and reflecting on what might be changed.

Relation to number
Knowing that the last number when counting units represents the measurement of the object.

Relationships
Connections between people that are based on mutual trust and respect.

Repeating patterns
Repeated sequences of items, which may be constructed using different layouts.

Representation theory
A theory focused on the construction of representations and the contexts that influence these representations.

Representations
Images, signs, characters, or objects that stand for or symbolise something.

Responding
Use of statements or questions that show understanding or invite responses.

Responsive strategies
Teaching practices that provide access to mathematical knowledge and generate respect, recognition, and cooperation.

Restrictive strategies
Teaching practices that restrict access to mathematical knowledge and create inequitable control and participation.

Rotational symmetry
The correspondence achieved between object and image when the image is turned through a fraction of a full circle.

Rote counting
Memorising and reciting numeral names and sequences.

Scaffolding
The process of building on a child's existing knowledge to introduce more complex knowledge.

Self-regulation
The capacity to use thought to guide behaviour.

Semi-structured number line
A number line with some of the marks for numbers.

Sensorimotor period
Understanding the present and the real world.

Sequence
The order in which events occur.

Seriation
The ability to sequence based on specific attributes.

Shared contexts
Contexts that consist of equity, balance, and flexibility in participation, meaning-making, and knowledge production.

Skip counting
Counting in multiples of any number other than one.

Social constructivism
A theory based on the belief that cognitive development results from interactions with the social and cultural world.

Social-conventional knowledge
Information gained through direct social transmission.

Social justice
Valuing diversity and providing equal opportunities.

Socio-economic status (SES)
A measure of an individual or family's economic or social position in relation to others.

Space
Understanding of the properties of objects as well as the relationships between objects.

Spatial orientation
Knowledge of one's place within an environment and the ability to relate meaningfully to objects within it.

Spatial reasoning
The ability to see, inspect, and reflect on spatial objects, images, relationships, and transformations.

Spatial visualisation
The ability to generate and manipulate a mental image or representation.

Stable order principle
Realisation that the counting sequence stays consistent.

Statistical literacy
The ability to make critical judgments about the ways in which data are presented and used.

Statistics
The processes of collecting, recording, describing, displaying, and organising data.

Strengths
Intellectual, physical and interpersonal skills, capacities, interests, and motivations.

Strengths approach
A way of working with children and families to help build their resilience.

Strengths-based learning plans
Structured planning documents that scaffold planning on the basis of children's strengths and interests.

Strengths-based practice
Educational practice that recognises and utilises children's strengths.

Structure
The features and characteristics of a learning domain, such as mathematics.

Structured number line
A number line marked with all the marks for numbers.

Subitising
The process of recognising how many items are in a small group.

Subtraction
Removing a part from a whole to find the remaining part.

Superimpose
Laying an object on top of another object.

Symbolic representations
Use of abstract symbols to depict relationships.

Symbolism
Representation of things with symbols, including language.

Symmetrical patterns
Patterns formed on the basis of the reflective or rotational correspondence of items.

Symmetry
Exactly similar parts facing each other or assembled around an axis.

Techno-toys
Toys that incorporate technologies—such as embedded electronics, response systems, and microchips—in their design.

Temperature
The warmth or coldness of an object or substance.

Tessellation
Fitting shapes together without gaps or overlaps.

Three-dimensional (3D) shapes
Solid shapes that have length, width, and height.

Time
The notion that events occur in a temporal order, and that events have duration.

Topological transformations
Alterations involving stretching or bending.

Transformation
The alteration of a shape.

Transition statements
Summaries of children's prior-to-school learning.

Transition to school
The movement of a child from their prior-to-school setting to primary school.

Transitivity
Knowing that if *A* is as long as *B*, and *B* is as long as *C*, then *A* is the same length as *C*.

Two-dimensional (2D) shapes
Flat shapes that have length and width but no height.

Understanding (mathematical)
Possession of a robust knowledge of adaptable and transferable mathematical concepts.

Unit
A quantity used as a standard of measurement.

Unit iteration
Thinking of the length of a unit as part of the length of an object, and placing that unit end to end along the object.

Value
The measurement of an object's importance, worth, or usefulness.

Volume
The amount of space taken up by an object.

Weight
The force that gravity exerts on an object.

Whole child
The recognition that each child is a unique individual who is shaped by many influences.

Whole numbers
Positive numbers, including zero.

Width
How wide something is, along a single plane or through two dimensions.

Working mathematically
Using the processes of mathematics.

Zone of Proximal Development (ZPD)
The difference between what a child can learn with and without assistance.

Bibliography

Aitken, J., Hunt, J., Roy, E., & Sajfar, B. (2012). *A sense of wonder: Science in early childhood education*. Albert Park, VIC: Teaching Solutions.

Antell, S.E., & Keating, D.P. (1983). Perception of numerical invariance in neonates. *Child Development, 54*, 695–701.

Arthur, L., Beecher, B., Death, E., Dockett, S., & Farmer, S. (2012). Contemporary perspectives on children's learning, development and play. In *Programming and planning in early childhood settings* (5th ed., pp. 72–107). South Melbourne, VIC: Cengage Learning Australia.

Arthur, L., Beecher, B., Death, E., Dockett, S., & Farmer, S. (2015). Thinking about children: Play, learning and development. In *Programming and planning in early childhood settings* (6th ed., pp. 69–106). South Melbourne, VIC: Cengage Learning Australia.

Atweh, B., Vale, C., & Walshaw, M. (2012). Equity, diversity, social justice and ethics in mathematics education. In B. Perry, T. Lowrie, T. Logan, A. MacDonald, & J. Greenelees (eds), *Research in mathematics education in Australasia 2008–2011* (pp. 39–65). Rotterdam, The Netherlands: Sense Publishers.

Aubrey, C. (1993). An investigation of the mathematical knowledge and competencies which young children bring into school. *British Educational Research Journal, 19*(1), 27–41.

Australian Association of Mathematics Teachers. (2006). *Standards for excellence in teaching mathematics in Australian schools*. Adelaide: AAMT.

Australian Association of Mathematics Teachers, & Early Childhood Australia. (2006). *Position paper on early childhood mathematics*. Adelaide, SA & Watson, ACT: Authors.

Australian Children's Education and Care Quality Authority. (2013). *Guide to the National Quality Standard*. Retrieved from: http://files.acecqa.gov.au/files/National-Quality-Framework-Resources-Kit/NQF03-Guide-to-NQS-130902.pdf

Australian Curriculum, Assessment and Reporting Authority (ACARA). (2022). *Australian Curriculum: Mathematics (Version 8.4)*. Available online: https://www.australiancurriculum.edu.au/f-10-curriculum/mathematics/

Australian Government Department of Education and Training (DET). (2019). *Belonging, being and becoming: The Early Years Learning Framework for Australia*. Available online: https://www.dese.gov.au/child-care-package/resources/belonging-being-becoming-early-years-learning-framework-australia

Bailey, J. (2014). Mathematical investigations for supporting pre-service primary teachers repeating a mathematics education course. *Australian Journal of Teacher Education, 39*(2), 86–100.

Balfanz, R. (1999). Why do we teach young children so little mathematics? Some historical considerations. In J.V. Copley (ed.), *Mathematics in the early years* (pp. 3–10). Resto, VA: National Council of Teachers of Mathematics.

Barnes, M.K. (2006). 'How many days 'til my birthday?' Helping kindergarten students understand calendar connections and concepts. *Teaching Children Mathematics, 12*(6), 290–5.

Baroody, A.J. (2000). Does mathematics for three and four year old children really make sense? *Young Children, 55*(4), 61–7.

Baroody, A.J., & Wilkins, J.L.M. (1999). The development of informal counting, number, and arithmetic skills and concepts. In J.V. Copley (ed.), *Mathematics in the early years* (pp. 48–65). Reston, VA: National Council of Teachers of Mathematics.

Basile, C.G. (1999). The outdoors as a context for mathematics in the early years. In J.V. Copley (ed.), *Mathematics in the early years* (pp. 156–61). Reston, VA: National Council of Teachers of Mathematics.

Battista, M.T. (2007). The development of geometric and spatial thinking. In F.K. Lester (ed.), *Second handbook of research on mathematics teaching and learning* (pp. 843–908). Reston, VA: National Council of Teachers of Mathematics.

Battista, M.T., & Clements, D.H. (1996). Students' understanding of three-dimensional rectangular arrays of cubes. *Journal for Research in Mathematics Education, 27*, 258–92.

Bennett, E., & Weidner, J. (2014). *The building blocks of early maths: Bringing key concepts to life for 3–6 year olds*. New York: Routledge.

Beswick, K. (2005). The beliefs/practice connection in broadly defined contexts. *Mathematics Education Research Journal, 17*(2), 39–68.

Bishop, A.J. (1988). *Mathematical enculturation: A cultural perspective on mathematics education*. Dordrecht, The Netherlands: Kluwer.

Bishop, A.J. (1991). *Mathematical enculturation: A cultural perspective on mathematics education*. Dordrecht, The Netherlands: Kluwer Academic Publishers.

Björklund, C. (2008). Toddlers' opportunities to learn mathematics. *International Journal of Early Childhood, 40*(1), 81–95.

Blair, C., Calkins, S., & Kopp, L. (2010). Self-regulation as the interface of emotional and cognitive development: Implications for education and academic achievement. In R.H. Hoyle (ed.), *Handbook of personality and self-regulation* (pp. 64–90). Oxford, UK: Wiley-Blackwell.

Boaler, J. (2013). Ability and mathematics: The mindset revolution that is reshaping education. *Forum, 55*(1), 143–52.

Bobis, J., Mulligan, J., & Lowrie, T. (2013). *Mathematics for children: Challenging children to think mathematically* (4th ed.). Frenchs Forest, NSW: Pearson Australia.

Booker, G., Bond, D., Sparrow, L., & Swan, P. (2014). *Teaching primary mathematics* (5th ed.). Frenchs Forest, NSW: Pearson Australia.

Boulton-Lewis, G. (1987). Recent cognitive theories applied to sequential length measuring knowledge in young children. *British Journal of Educational Psychology, 57*, 330–42.

Brady, K. (2008). Using paper-folding in the primary years to promote student engagement in mathematical learning. In M. Goos, R. Brown, & K. Makar (eds), *Navigating currents and charting directions: Proceedings of the 31st annual conference of the Mathematics Education Research Group of Australasia* (vol. 1, pp. 77–84). Brisbane: MERGA.

Bronfenbrenner, U. (1979). *The ecology of human development: Experiments by nature and design*. Cambridge, MA: Harvard University Press.

Bronfenbrenner, U. (1988). Interacting systems in human development. Research paradigms: Present and future. In N. Bolger, A. Caspi, G. Downey, & M. Moorehouse (eds), *Persons in context: Developmental processes* (pp. 25–49). Cambridge, MA: Cambridge University Press.

Brooks, M. (2004). Drawing: The social construction of knowledge. *Australian Journal of Early Childhood, 29*(2), 41–9.

Bruce, T. (2011). *Learning through play: For babies, toddlers and young children* (2nd ed.). London: Hodder Education.

Bruner, J. (ed.). (1966). *Studies in cognitive growth*. New York: Wiley and Son.

Bull, R., & Lee, K. (2014). Executive functioning and mathematics achievement. *Child Development Perspectives, 8*, 36–41.

Cairney, T. (2003). Literacy within family life. In N. Hall, J. Larson, & J. Marsh (eds), *Handbook of early childhood literacy* (pp. 85–98). London: Sage Publications.

Carmichael, C., MacDonald, A., & McFarland-Piazza, L. (2014). Predictors of numeracy performance in national testing programs: Insights from the Longitudinal Study of Australian Children. *British Educational Research Journal, 40*(4), 637–59.

Carr, M. (2001). *Assessment in early childhood settings: Learning stories*. London: Paul Chapman.

Carruthers, E., & Worthington, M. (2006). *Children's mathematics: Making marks, making meaning* (2nd ed.). London: SAGE.

Charlesworth, R. (2005). *Experiences in math for young children* (5th ed.). Clifton Park, NY: Thomson Delmar Learning.

Clark, A. (2005). Listening to and involving young children: A review of research and practice. *Early Child Development and Care, 175*(6), 489–505.

Clarke, B., Clarke, D., & Cheeseman, J. (2006). The mathematical knowledge and understanding young children bring to school. *Mathematics Education Research Journal, 18*(1), 78–102.

Clements, D.H. (1999a). Geometric and spatial thinking in young children. In. J.V. Copley (ed.), *Mathematics in the early years* (pp. 66–79). Reston, VA: National Council of Teachers of Mathematics.

Clements, D.H. (1999b). Teaching length measurement: Research challenges. *School Science and Mathematics, 99*(1), 5–11.

Clements, D.H., & Sarama, J. (2014). *Learning and teaching early math: The learning trajectories approach* (2nd ed.). New York: Routledge.

Clements, D.H., & Stephan, M. (2004). Measurement in pre-K to grade 2 mathematics. In D.H. Clements & J. Sarama (eds), *Engaging young children in mathematics: Standards for early childhood mathematics education* (pp. 299–320). Mahwah, NJ: Lawrence Erlbaum Associates, Inc.

Coates, G.D., & Thompson, V. (1999). Involving parents of four- and five-year-olds in their children's mathematics education: The FAMILY MATH experience. In J.V. Copley (ed.), *Mathematics in the early years* (pp. 205–14). Reston, VA: National Council of Teachers of Mathematics.

Collins, F., & Fenton, A. (2021). An introduction to the Strengths Approach. In Y.H. Leong, B. Kaur, B.H. Choy, J.B.W. Yeo, & S.L. Chin (eds), *Excellence in mathematics education: Foundations and pathways: Proceedings of the 43rd annual conference of the Mathematics Education Research Group of Australasia* (pp. 84–87). Singapore: MERGA.

Copley, J.V. (1999). Assessing the mathematical understanding of the young child. In J.V. Copley (ed.), *Mathematics in the early years* (pp. 182–8). Reston, VA: National Council of Teachers of Mathematics.

Copley, J.V. (2001). *The young child and mathematics*. Washington, DC: National Association for the Education of Young Children.

Copley, J.V. (2010). *The young child and mathematics* (2nd ed.). Reston, VA; National Council of Teachers of Mathematics.

Copley, J.V., Jones, C., & Dighe, J., with Bickart, T.S., & Herornan, C. (2007). *Mathematics: The creative curriculum approach*. Washington, DC: Teaching Strategies.

Curry, M., & Outhred, L. (2005). Conceptual understanding of spatial measurement. In P. Clarkson, A. Downton, D. Gronn, M. Horne, A. McDonough, R. Pierce, & A. Roche (eds), *Building connections: Theory, research and practice: Proceedings of the 27th annual conference of the Mathematics Education Research Group of Australasia* (pp. 265–72). Sydney: MERGA.

De Lange, J. (2008). *Talentenkracht* [Curious minds]. The Netherlands: Freudenthal Institute for Mathematics and Science Education.

Department of Education, Employment and Workplace Relations. (2006). *Early childhood learning* [DVD]. Barton, ACT: Commonwealth of Australia.

Department of Education, Employment and Workplace Relations. (2010). *Educators belonging, being and becoming: Educators' guide to the Early Years Learning Framework for Australia.* Barton, ACT: Commonwealth of Australia.

Department of Education, Employment and Workplace Relations. (2011). *My time, our place: Framework for School Age Care in Australia.* Barton, ACT: Commonwealth of Australia.

Department of Education and Training. (2019). *See* Australian Government Department of Education and Training

De Vries, E., Thomas, L., & Warren, E. (2010). Teaching mathematics and play-based learning in an indigenous early childhood setting: Early childhood teachers' perspectives. In L. Sparrow, B. Kissane, & C. Hurst (eds), *Shaping the future of mathematics education: Proceedings of the 33rd annual conference of the Mathematics Education Research Group of Australasia* (pp. 719–22). Fremantle: MERGA.

Dockett, S., & Perry, B. (2006). *Starting school: A handbook for early childhood educators*. Castle Hill, NSW: Pademelon Press.

Dockett, S., & Perry, B. (2007). *Transitions to school: Perceptions, experiences and expectations*. Sydney: University of New South Wales Press.

Dockett, S., & Perry, B. (2014). *Continuity of learning: A resource to support effective transition to school and school age care.* Canberra, ACT: Australian Government Department of Education.

Dockett, S., Perry, B., Kearney, E., Hampshire, A., Mason, J., & Schmied, V. (2011). *Facilitating children's transition to school from families with complex support needs*. Albury, NSW: Research Institute for Professional Practice, Learning and Education, Charles Sturt University.

Doig, B., McRae, B., & Rowe, K. (2003). *A good start to numeracy: Effective numeracy strategies from research and practice in early childhood*. Canberra: Commonwealth Department of Education, Science and Training.

Duncan, G.J., Dowsett, C.J., Claessens, A., Magnuson, K., Huston, A.C., Klebanov, P., ... Japel, C. (2007). School readiness and later achievement. *Developmental Psychology, 43*(6), 1428–46.

Dunphy, L. (2020). A picture book pedagogy for early childhood mathematics education. In A. MacDonald, L. Danaia, & S. Murphy (eds), *STEM education across the learning continuum: Early childhood to senior secondary* (pp. 67–85). Singapore: Springer.

Dweck, C.S. (2006). *Mindset: The new psychology of success.* New York: Ballantine Books.

Ebbeck, M., & Waniganayake, M. (2010). Perspectives on play in a changing world. In M. Ebbeck & M. Waniganayake (eds), *Play in early childhood education: Learning in diverse contexts* (pp. 5–25). South Melbourne, VIC: Oxford University Press.

Educational Transitions and Change (ETC) Research Group. (2011). *Transition to school: Position statement.* Albury–Wodonga, Australia: Research Institute for Professional Practice, Learning and Education (RIPPLE), Charles Sturt University. Retrieved from: https://arts-ed.csu.edu.au/education/transitions/publications/Position-Statement.pdf

Einarsdóttir, J. (2007). Research with children: Methodological and ethical challenges. *European Early Childhood Education Research Journal, 15*(2), 197–211.

Ernest, P. (1989). The impact of beliefs on the teaching of mathematics. In P. Ernest (ed.), *Mathematics teaching: The state of the art* (pp. 249–53). New York: Falmer.

Fenton, A., MacDonald, A., & McFarland-Piazza, L. (2016). A Strengths Approach to supporting early mathematics learning in family contexts. *Australasian Journal of Early Childhood, 41*(1), 45–53.

Fenton, A., & McFarland-Piazza, L. (2014). Supporting early childhood preservice teachers in their work with children and families with complex needs: A Strengths Approach. *Journal of Early Childhood Teacher Education, 35*(1), 22–38.

Fox, M., & Horacek, J. (2004). *Where is the green sheep?* Melbourne: Puffin Books.

Garbarino, J., & Plantz, M.C. (1980). *Urban environments and urban children.* ERIC Clearinghouse on Urban Education.

Geary, D.C. (1994). Children's early numerical abilities. In *Children's mathematical development: Research and practical applications* (pp. 1–35). Washington, DC: American Psychological Association.

Geist, E. (2009). *Children are born mathematicians: Supporting mathematical development, birth to age 8.* Upper Saddle River, NJ: Pearson Education.

Gervasoni, A., & Perry, B. (2012). *Let's Count facilitator's guide.* Sydney: The Smith Family.

Gervasoni, A., & Perry, B. (2015). Children's mathematical knowledge prior to starting school and implications for transition. In B. Perry, A. MacDonald, & A. Gervasoni (eds), *Mathematics and transitions to school: International perspectives* (pp. 47–64). Dordrecht, The Netherlands: Springer.

Gifford, S. (2005). *Teaching mathematics 3–5: Developing learning in the foundation stage.* Maidenhead, England: Open University Press.

Ginsburg, H.P., Inoue, N., & Seo, K. (1999). Young children doing mathematics: Observations of everyday activities. In J.V. Copley (ed.), *Mathematics in the early years* (pp. 88–99). Reston, VA: National Council of Teachers of Mathematics.

Goff, W. (2016). Partnership at the cultural interface: How adults come together to support the mathematical learning of children making the transition to school. Unpublished PhD thesis, Charles Sturt University.

Goldin, G.A., & Kaput, J.J. (1996). A joint perspective on the idea of representation in learning and doing mathematics. In L.P. Steffe, P. Nesher, P. Cobb, G.A. Goldin, & B. Greer (eds), *Theories of mathematical learning* (pp. 397–430). Mahwah, NJ: Lawrence Erlbaum Associates, Inc.

Goldin, G.A., & Shteingold, N. (2001). Systems of representations and the development of mathematical concepts. In A.A. Cuoco & F.R. Curcio (eds), *The roles of representation in school mathematics* (pp. 1–23). Reston, VA: National Council of Teachers of Mathematics.

Goodwin, K., & Highfield, K. (2012). iTouch and iLearn: An examination of 'educational' Apps. Paper presented at the Early Education and Technology for Children Conference, Salt Lake City, USA, 14–16 March.

Goodwin, K., & Highfield, K. (2013). A framework for examining technologies and early mathematics learning. In L.D. English & J.T. Mulligan (eds), *Reconceptualising early mathematics learning* (pp. 205–26). Dordrecht, The Netherlands: Springer.

Goos, M., Lowrie, T., & Jolly, L. (2007). Home, school and community partnerships in numeracy education: An Australian perspective. *The Montana Mathematics Enthusiast, 1,* 7–24.

Gould, P. (2012). What number knowledge do children have when starting kindergarten in NSW? *Australasian Journal of Early Childhood, 37*(3), 105–10.

Gould, P. (2014). The association between students' number knowledge and social disadvantage at school entry. In J. Anderson, M. Cavanagh, & A. Prescott (eds), *Curriculum in focus: Research guided practice: Proceedings of the 37th annual conference of the Mathematics Education Research Group of Australasia* (pp. 255–62). Sydney: MERGA.

Graue, E., Karabon, A., Delaney, K.K. Whyte, K., Kim, J., & Wager, A. (2015). Imagining a future in PreK: How professional identity shapes notions of early mathematics. *Anthropology and Education Quarterly, 46*(1), 37–54.

Greenes, C. (1999). Ready to learn: Developing young children's mathematical powers. In J.V. Copley (ed.), *Mathematics in the early years* (pp. 39–47). Reston, VA: National Council of Teachers of Mathematics.

Grootenboer, P., & Marshman, M. (2016). The affective domain, mathematics, and mathematics education. In *Mathematics, affect and learning* (pp. 13–33). Singapore: Springer.

Gutiérrez, A. (1996). Visualization in 3-dimensional geometry: In search of a framework. In L. Puig & A. Gutiérrez (eds), *Proceedings of the 20th Conference of the International Group for the Psychology of Mathematics Education* (vol. 1, pp. 3–19). Valencia: Universidad de Valencia.

Harlan, J.D., & Rivkin, M.S. (2012). *Science experiences for the early childhood years: An integrated affective approach* (10th ed.). Upper Saddle River, NJ: Pearson Education.

Hiebert, J. (1981). Cognitive development and learning linear measurement. *Journal for Research in Mathematics Education, 12*(3), 197–211.

Highfield, K. (2010). Possibilities and pitfalls of techno-toys and digital play in mathematics learning. In M. Ebbeck & M. Waniganayake (eds), *Play in early childhood education: Learning in diverse contexts* (pp. 177–96). South Melbourne, VIC: Oxford University Press.

Hilton, G., Dole, S., & Campbell, C. (2014). *Teaching early years mathematics, science and ICT.* Sydney: Alllen & Unwin.

Hong, H. (1999). Using storybooks to help young children make sense of mathematics. In J.V. Copley (ed.), *Mathematics in the early years* (pp. 162–8). Reston, VA: National Council of Teachers of Mathematics.

Howard, P.T. (2001). *Beliefs about the nature and learning of mathematics in Years 5 and 6: The voices of Aboriginal children, parents, Aboriginal educators and teachers.* Unpublished PhD thesis, University of Western Sydney.

Howell, J., & McMaster, N. (2022). *Teaching with Technologies: Pedagogies for collaboration, communication and creativity.* Melbourne, VIC: Oxford University Press.

Hunter, J., & Miller, J. (2022). Using a culturally responsive approach to develop early algebraic reasoning with young diverse learners. *International Journal of Science and Mathematics Education, 20,* 111–31.

Hunting, R., Mousley, J., & Perry, B. (2012). *Young children learning mathematics: A guide for educators and families.* Camberwell, VIC: ACER Press.

Ivrendi, A. (2011). Influence of self-regulation on the development of children's number sense. *Early Childhood Education Journal, 39,* 239–47.

Izsák, A. (2005). 'You have to count the squares': Applying knowledge in pieces to learning rectangular area. *Journal of the Learning Sciences, 14*(3), 361–403.

Jorgensen, R., & Dole, S. (2011). *Teaching mathematics in primary schools* (2nd ed.). Crows Nest, NSW: Allen & Unwin.

Kamii, C., & Clark, F. (1997). Measurement of length: The need for a better approach to teaching. *School Science and Mathematics, 97*(3), 116–21.

Kearns, K. (2010). *Birth to big school.* Frenchs Forest, NSW: Pearson Australia.

Kilpatrick, J., Swafford, J., & Findell, B. (eds) (2001). *Adding it up: Helping children learn mathematics.* Washington, DC: National Academy Press.

Kim, S.L. (1999). Teaching mathematics through musical activities. In J.V. Copley (ed.), *Mathematics in the early years* (pp. 146–50). Reston, VA: National Council of Teachers of Mathematics.

Klibanoff, R. (2006). Preschool children's mathematical knowledge: The effect of teacher 'math talk'. *Developmental Psychology, 42*(1), 59–69.

Knaus, M. (2013). *Maths is all around you: Developing mathematical concepts in the early years.* Albert Park, VIC: Teaching Solutions.

Lee, S. (2012). Toddlers as mathematicians? *Australasian Journal of Early Childhood, 37*(1), 30–37.

Lee, P. (2022). *Investigating mathematics with babies and toddlers in informal settings in Australia.* Unpublished MEd thesis, Charles Sturt University.

Lehrer, R. (2003). Developing understanding of measurement. In J. Kilpatrick, W.G. Martin, & D. Shifter (eds), *A research companion to principles and standards for school mathematics* (pp. 179–93). Reston, VA: National Council of Teachers of Mathematics.

Lehrer, R., Jacobson, C., Kemeny, V., & Strom, D. (1999). Building on children's intuitions to develop mathematical understanding of space. In E. Fennema & T.A. Romberg (eds), *Mathematics classrooms that promote understanding* (pp. 63–88). Mahwah, NJ: Lawrence Erlbaum Associates, Publishers.

Lehrer, R., Jenkins, M., & Osana, H. (1998). Longitudinal study of children's reasoning about space and geometry. In R. Lehrer & D. Chazan (eds), *Designing learning environments for developing understandings of geometry and space* (pp. 137–67). Mahwah, NJ: Lawrence Erlbaum Associates, Inc.

'Let the children play'. (2010). Retrieved from: http://www.letthechildrenplay.net

Levine, S.C., Suriyakham, L.W., Rowe, M.L., Huttenlocher, J., & Gunderson, E.A. (2010). What counts in the development of young children's number knowledge? *Developmental Psychology, 46*(5), 1309–19.

Liebeck, P. (1984). *How children learn mathematics.* London: Penguin.

Lind, K.K. (1998). Science in early childhood: Developing and acquiring fundamental concepts and skills. Paper presented at the Forum on Early Childhood Science, Mathematics, and Technology Education, Washington DC, 6–8 February.

Linder, S.M., Powers-Costello, B., & Stegelin, D.A. (2011). Mathematics in early childhood: Research-based rationale and practical strategies. *Early Childhood Education Journal, 39,* 29–37.

Lowrie, T., Logan, T., & Ramful, A. (2016). Spatial reasoning influences students' performance on mathematics tasks. In B. White, M. Chinnappan, & S. Trenholm (eds), *Opening up mathematics education research: Proceedings of the 39th annual conference of the Mathematics Education Research Group of Australasia* (pp. 407–14). Adelaide: MERGA.

Ma, X. (1997). Reciprocal relationships between attitude toward mathematics and achievement in mathematics. *Journal of Educational Research, 90*(4), 221–9.

McCashen, W. (2005). *The strengths approach: A strengths-based resource for sharing power and creating change.* Bendigo, VIC: St Luke's Innovative Resources.

MacDonald, A. (2010). Young children's measurement knowledge: Understandings about comparison at the commencement of schooling. In L. Sparrow, B. Kissane, & C. Hurst (eds), *Shaping the future of mathematics education: Proceedings of the 33rd annual conference of the Mathematics Education Research Group of Australasia* (pp. 375–82). Fremantle: MERGA.

MacDonald, A. (2013). Using children's representations to investigate meaning-making in mathematics. *Australasian Journal of Early Childhood, 38*(2), 65–73.

MacDonald, A. (2015a). *Investigating mathematics, science and technology in early childhood.* South Melbourne, VIC: Oxford University Press.

MacDonald, A. (2015b). *Let's Count*: Early childhood educators and families working in partnership to support young children's transitions in mathematics education. In B. Perry, A. MacDonald, & A. Gervasoni (eds), *Mathematics and transition to school: International perspectives* (pp. 85–102). Singapore: Springer.

MacDonald, A., & Carmichael, C. (2015). A snapshot of young children's mathematical competencies: Results from the Longitudinal Study of Australian Children. In M. Marshman, V. Geiger, & A. Bennison (eds), *Mathematics education in the margins: Proceedings of the 38th Annual Conference of the Mathematics Education Research Group of Australasia* (pp. 381–8). Sunshine Coast, QLD: MERGA.

MacDonald, A., & Carmichael, C. (2016). Early mathematical competencies and later outcomes: Insights from the Longitudinal Study of Australian Children. In B. White, M. Chinnappan, & S. Trenholm (eds), *Opening up mathematics education research: Proceedings of the 39th annual conference of the Mathematics Education Research Group of Australasia* (pp. 413–20). Adelaide: MERGA.

MacDonald, A., Goff, W., Dockett, S., & Perry, B. (2016). Mathematics education in the early years. In K. Makar, S. Dole, J. Visnovska, M. Goos, A. Bennison, & K. Fry (eds), *Research in mathematics education in Australasia 2012–2015* (pp. 165–86). Singapore: Springer.

MacDonald, A., & McGrath, S. (2022). Early childhood educators' beliefs about mathematics education for children under three years of age. *International Journal of Early Years Education*. https://doi.org/10.1080/09669760.2022.2107493

MacGregor, M., & Stacey, K. (1999). A flying start to algebra. *Teaching Children Mathematics*, *6*(2), 78–85.

Macmillan, A. (1995). Children thinking mathematically beyond authoritative identities. *Mathematics Education Research Journal*, *7*(2), 111–31.

Macmillan, A. (2009). *Numeracy in early childhood: Shared contexts for teaching and learning*. South Melbourne, VIC: Oxford University Press.

McPhail, D. (2007). Area: The big cover-up. Unpublished PhD thesis, University of Western Sydney.

Mallucio, A.N. (1981). *Promoting competence in clients: A new/old approach to social work practice*. New York: The Free Press.

Maranhãa, C., & Campos, T. (2000). Length measurement: Conventional units articulated with arbitrary ones. In T. Nakahara & M. Koyama (eds), *Proceedings of the 24th annual conference of the International Group for the Psychology of Mathematics Education* (vol. 3, pp. 255–62). Hiroshima, Japan: PME.

Marcus, A., Perry, B., Dockett, S., & MacDonald, A. (2016). Children noticing their own and others' mathematics in play. In B. White, M. Chinnappan, & S. Trenholm (eds), *Opening up mathematics education research: Proceedings of the 39th annual conference of the Mathematics Education Research Group of Australasia* (pp. 437–44). Adelaide: MERGA.

Mason, J. (2002). *Researching your own practice: The discipline of noticing*. London: RoutledgeFalmer.

Maxwell, L.E., Mitchell, M.R., & Evans, G.W. (2008). Effects of play equipment and loose parts on preschool children's outdoor play behaviour: An observational study and design intervention. *Children, Youth and Environments*, *18*(2), 36–63.

Meaney, T., & Lange, T. (2011). *Mathematics: Content and pedagogy* (EMM209 modules). Wagga Wagga, NSW: Charles Sturt University.

Miller, J. (2019). STEM education in the primary years to support mathematical thinking: Using coding to identify mathematical structures and patterns. *ZDM*, *51*(6), 915–27.

Montague-Smith, A., & Price, A.J. (2012). *Mathematics in early years education* (3rd ed.). New York: Routledge.

Mulligan, J., Mitchelmore, M., & Prescott, A. (2006). Integrating concepts and processes in early mathematics. In J. Novotna, H. Moraova, M. Kratka, & N. Stehlikova (eds), *Proceedings of the 30th annual conference of the International Group for the Psychology of Mathematics Education* (vol. 4, pp. 209–16). Prague: PME.

Murphy, S., MacDonald, A., Wang, C., & Danaia, L. (2019). Towards an understanding of STEM engagement: A review of the literature on motivation and academic emotions. *Canadian Journal of Science, Mathematics and Technology Education*, *19*, 304–20.

National Council of Teachers of Mathematics. (1995). *Assessment standards for school mathematics*. Reston, VA: Author.

National Council of Teachers of Mathematics. (2000). *Principles and standards for school mathematics*. Reston, VA: Author.

National Curriculum Board. (2009). *Shape of the Australian Curriculum: Mathematics*. Available online: https://docs.acara.edu.au/resources/Australian_Curriculum_-_Maths.pdf

New South Wales Department of Education and Communities. (2017). NSW Transition to School Statement. Retrieved from: https://www.det.nsw.edu.au/media/downloads/what-we-offer/regulation-and-accreditation/early-childhood-education-care/funding/funding-projects/tts/TTS-Statement-for-completion.pdf

Nicolaou, C., Evagorou, M., & Lymbouridou, C. (2015). Elementary school students' emotions when exploring an authentic socio-scientific issue through the use of models. *Science Education International*, *26*(2), 240–59.

Organisation for Economic Co-operation and Development. (2013). *PISA 2012 assessment and analytical framework: Mathematics, reading, science, problem solving and financial literacy*. Paris: OECD Publishing.

Owens, K., McPhail, D., & Reddacliff, C. (2003). Facilitating the teaching of space mathematics: An evaluation. In N. Pateman, B. Dougherty, & J. Zilliox (eds), *Proceedings of 27th annual conference of the International Group for the Psychology of Mathematics Education* (vol. 1, pp. 339–45). Hawaii: International Group for the Psychology of Mathematics Education.

Pajares, M.F. (1992). Teachers' beliefs and educational research: Cleaning up a messy construct. *Review of Educational Research*, *62*(3), 307–32.

Papic, M. (2007). *Mathematical patterning in early childhood: An intervention study.* Unpublished PhD thesis, Macquarie University.

Pekrun, R., Goetz, T., Titz,W., & Perry, R.P. (2002). Academic emotions in students' self-regulated learning and achievement: A program of qualitative and quantitative research. *Educational Psychologist*, *37*(2), 91–105.

Perry, B., & Conroy, J. (1994). *Early childhood and primary mathematics*. Sydney: Harcourt Brace.

Perry, B., & Dockett, S. (2004). Mathematics in early childhood education. In B. Perry, G. Anthony, & C. Diezmann (eds), *Review of mathematics education research in Australasia: 2000–2003* (pp. 103–25). Brisbane: Post Pressed.

Perry, B., & Dockett, S. (2005a). 'I know that you don't have to work hard': Mathematics learning in the first year of primary school. In H. Chick & J.L. Vincent (eds), *Proceedings of the 29th conference of the International Group for the Psychology of Mathematics Education* (vol. 4, pp. 65–72). Melbourne: PME.

Perry, B., & Dockett, S. (2005b). What did you do in maths today? *Australian Journal of Early Childhood*, *30*(3), 32–36.

Perry, B., & Dockett, S. (2008). Young children's access to powerful mathematical ideas. In L.D. English (ed.), *Handbook of international research in mathematics education* (2nd ed., pp. 75–108). New York: Routledge.

Perry, B., Dockett, S., & Harley, E. (2012). The Early Years Learning Framework for Australia and the Australian Curriculum: Mathematics—Linking educators' practice through pedagogical inquiry questions. In B. Atweh, M. Goos, R. Jorgensen, & D. Siemon (eds), *Engaging the Australian National Curriculum: Mathematics—Perspectives from the field* (pp. 155–74). Mathematics Education Research Group of Australasia. Available online: https://www.merga.net.au/common/Uploaded%20files/Publications/Engaging%20the%20Australian%20Curriculum%20Mathematics.pdf

Perry, B., & Gervasoni, A. (2012). *Let's Count educators' handbook*. Sydney: The Smith Family.

Perry, B., MacDonald, A., & Gervasoni, A. (2015). Mathematics and transition to school: Theoretical frameworks and practical implications. In B. Perry, A. MacDonald, & A. Gervasoni (eds), *Mathematics and transition to school: International perspectives* (pp. 1–12). Singapore: Springer.

Peter-Koop, A., & Kollhoff, S. (2015). Transition to school: Prior to school mathematical skills and knowledge of low-achieving children at the end of Grade 1. In B. Perry, A. MacDonald, & A. Gervasoni (eds), *Mathematics and transition to school: International perspectives* (pp. 65–83). Singapore: Springer.

Philipp, R.A. (2007). Mathematics teachers' beliefs and affect. In F.K. Lester (ed.), *Second handbook of research on mathematics teaching and learning* (vol. 1). Reston, VA: National Council of Teachers of Mathematics.

Piaget, J. (1969a). *Science of education and the psychology of the child*. New York: Viking Press.

Piaget, J. (1969b). *The child's conception of time*. London: Routledge & Kegan Paul.

Pirie, S., & Kieren, T. (1994). Growth in mathematical understanding: How can we characterise it and how can we represent it? *Educational Studies in Mathematics, 26*(2 and 3), 165–90.

Queensland Department of Education and Training. (2013). Transition to school: Summary of teacher information [sample]. Retrieved from: https://www.qcaa.qld.edu.au/downloads/p_10/qklg_pd_transition_statement_sample_1.pdf

Raising Children Network. (2017). Developing early numeracy skills. Retrieved from: http://raisingchildren.net.au/articles/developing_early_numeracy_skills.html

Reikerås, E., Løge, I.K., & Knivsberg, A. (2012). The mathematical competencies of toddlers expressed in their play and daily life activities in Norwegian kindergartens. *International Journal of Early Childhood, 44*, 91–114.

Reys, R.E., Lindquist, M.M., Lambdin, D.V., & Smith, N.L. (2007). *Helping children learn mathematics* (8th ed.). Hoboken, NJ: John Wiley & Sons, Inc.

Reys, R., Lindquist, M.M., Lambdin, D.V., & Smith, N.L. (2012). *Helping children learn mathematics* (10th ed.). Hoboken, NJ: John Wiley & Sons, Inc.

Rogoff, B. (1990). *Apprenticeship in thinking: Cognitive development in social context*. New York: Oxford University Press.

Rogoff, B. (2003). *The cultural nature of human development*. New York: Oxford University Press.

Sarama, J., & Clements, D. (2002). Building blocks for young children's mathematical development. *Journal of Educational Computing Research, 27*(1–2), 93–110.

Sarama, J., & Clements, D.H. (2009). *Early childhood mathematics educational research: Learning trajectories for young children*. London: Routledge.

Sarama, J., & Clements, D.H. (2015). Scaling up early childhood mathematics interventions: Transitioning with trajectories and technologies. In B. Perry, A. MacDonald, & A. Gervasoni (eds), *Mathematics and transition to school: International perspectives* (pp. 153–69). Dordrecht, The Netherlands: Springer.

Siegler, R.S., & Shrager, J. (1984). Strategy choice in addition and subtraction: How do children know what to do? In C. Sophian (ed.), *Origins of cognitive skills* (pp. 229–93). Hillsdale, NJ: Erlbaum.

Siemon, D., Beswick, K., Brady, K., Clark, J., Faragher, R., & Warren, E. (2011). *Teaching mathematics: Foundations to middle years*. South Melbourne, VIC: Oxford University Press.

Siemon, D., Warren, E., Beswick, K., Faragher, R., Miller, J., Horne, M., Jazby, D., & Breed, M. (2021). *Teaching mathematics: Foundations to middle years*. South Melbourne, VIC: Oxford University Press.

Simon, R.A., Aulls, M.W., Dedic, H., Hubbard, K., & Hall, N.C. (2015). Exploring student persistence in STEM programs: A motivational model. *Canadian Journal of Education, 38*(1).

Skinner, P. (1990). *What's your problem?* Portsmouth, NH: Heinemann.

Skwarchuk, S., & LeFevre, J. (2015). The role of the home environment in children's early numeracy development: A Canadian perspective. In B. Perry, A. MacDonald, & A. Gervasoni (eds), *Mathematics and transition to school: International perspectives* (pp. 103–17). Singapore: Springer.

Smith, T., & MacDonald, A. (2009). Time for talk: The drawing-telling process. *Australian Primary Mathematics Classroom, 14*(3), 21–6.

Southwell, B., & Khamis, M. (1994). *Beliefs of secondary school students concerning mathematics and mathematics education*. Paper presented at the annual conference of the Australian Association for Research in Education, Newcastle.

Stelzer, E. (2005). *Experiencing science and math in early childhood*. Toronto, ON: Pearson Canada.

Stephan, M., & Clements, D.H. (2003). Linear and area measurement in prekindergarten to grade 2. In D.H. Clements & G. Bright (eds), *Learning and teaching measurement*. NCTM 2003 Yearbook (pp. 3–16). Reston, VA: National Council of Teachers of Mathematics.

Stephan, M., & Cobb, P. (1998). The evolution of mathematical practices: How one first-grade classroom learned to measure. In A. Olivier & K. Newstead (eds), *Proceedings of the 22nd annual conference of the International Group for the Psychology of Mathematics Education* (vol. 4, pp. 97–104). Stellenbosch, South Africa: PME.

Tymms, P., Merrell, C., & Henderson, B. (1997). The first year at school: A quantitative investigation of the attainment and progress of pupils. *Educational Research and Evaluation, 3*(2), 101–18.

Vale, C., Atweh, B., Averill, R., & Skourdoumbis, A. (2016). Equity, social justice and ethics in mathematics education. In K. Makar, S. Dole, J. Visnovska, M. Goos, A. Bennison, & K. Fry (eds), *Research in mathematics education in Australasia 2012–2015* (pp. 97–118). Singapore: Springer.

van den Heuvel-Panhuizen, M., & Elia, H. (2012). Developing a framework for the evaluation of picturebooks that support kindergartners' learning of mathematics. *Research in Mathematics Education, 14*(1), 17–47.

Van Zoest, L.R., Jones, G.A., & Thornton, C.A. (1994). Beliefs about mathematics teaching held by pre-service teachers involved in a first grade mentorship program. *Mathematics Education Research Journal, 6*(1), 37–55.

Victorian Department of Education and Training. (2017). Transition Learning and Development Statement. Retrieved from: http://www.education.vic.gov.au/childhood/professionals/learning/Pages/transitionstat.aspx

Vygotsky, L. (1978). *Mind in society: The development of higher psychological processes*. Cambridge, MA: Harvard University Press.

Warren, E., Miller, J., & Cooper, T. (2012). Repeating patterns: Strategies to assist young students to generalise the mathematical structure. *Australasian Journal of Early Childhood, 37*(3), 111–20.

Warren, E., Young, J., & De Vries, E. (2008). Indigenous students' early engagement with numeracy: The case of Widgy and Caddy. In M. Goos, R. Brown, & K. Makar (eds), *Navigating currents and charting directions: Proceedings of the 31st annual conference of the Mathematics Education Research Group of Australasia* (pp. 547–54). Brisbane: MERGA.

Watts, T., Duncan, G., Siegler, R., & Davis-Kean, P. (2014). What's past is prologue: Relations between early mathematics knowledge and high school achievement. *Educational Researcher, 43*(7), 352–60.

Way, J., Bobis, J., Lamb, J., & Higgins, J. (2016). Researching curriculum, policy and leadership in mathematics education. In K. Makar, S. Dole, J. Visnovska, M. Goos, A. Bennison, & K. Fry (eds), *Research in mathematics education in Australasia 2012–2015* (pp. 49–71). Singapore: Springer.

Whitebread, D. (2005). Emergent mathematics or how to help young children become confident mathematicians. In J. Anghileri (ed.), *Children's mathematical thinking in the primary years: Perspectives on children's learning* (pp. 11–40). London: Continuum.

Williams, K.E., White, S.L.J., & MacDonald, A. (2016). Early mathematics achievement of boys and girls: Do differences in early self-regulation pathways explain later achievement? *Learning and Individual Differences, 51,* 199–209.

Wilson, S., & Raven, M. (2014). 'Change my thinking patterns towards maths': A bibliotherapy workshop for pre-service teachers' mathematics anxiety. In J. Anderson, M. Cavanagh, & A. Prescott (eds), *Curriculum in focus: Research guided practice: Proceedings of the 37th annual conference of the Mathematics Education Research Group of Australasia* (pp. 645–52). Sydney: MERGA.

Woleck, K.R. (2001). Listen to their pictures: An investigation of children's mathematical drawings. In A.A. Cuoco & F.R. Curcio (eds), *The roles of representation in school mathematics* (pp. 215–27). Reston, VA: National Council of Teachers of Mathematics.

Wright, R.J. (1994). A study of the numerical development of 5-year-olds and 6-year-olds. *Educational Studies in Mathematics, 26,* 25–44.

Index